Instrument Oral Exam Guide

The comprehensive guide to prepare you for the FAA Oral Exam

by Michael D. Hayes

Published by
Aviation Supplies & Academics, Inc.
Renton, WA 98059-3153

Instrument Oral Exam Guide

Aviation Supplies & Academics, Inc.
7005 132nd Place SE
Renton, Washington 98059-3153

98 97 96 95 94 93 92 9 8 7 6 5 4 3 2 1

ASA-OEG-I

This guide is dedicated to the many talented students, pilots and flight instructors I have had the opportunity to work with over the years. Also, special thanks to Mark Hayes and many others who supplied the patience, encouragement, and understanding necessary to complete the project.

CONTENTS

4. ARRIVAL

Introduction

The **COMPLETE GUIDE TO THE FAA INSTRUMENT ORAL** is a comprehensive guide designed for private or commercial pilots who are involved in training for the instrument rating. This guide was originally designed for use in a part 141 flight school but has quickly become popular with those training under FAR Part 61 who are not affiliated with an approved school. The guide will also prove beneficial to instrument rated pilots who wish to refresh their knowledge or who are preparing for an instrument competency check.

The guide is divided into four main parts which represent logical divisions of a typical instrument flight. During the instrument oral, an FAA examiner may ask questions pertaining to any of the subject areas within these divisions. Through very intensive post instrument check ride de-briefings, we have provided you with the most consistent questions asked along with the information necessary for a knowledgable response.

The guide may be supplemented with other comprehensive study materials listed below.

1. The Instrument Flying Handbook (AC61-27C)
2. Airmen's Information Manual (AIM)
3. Pilot's Handbook of Aeronautical Knowledge (AC61-23B)
4. Aviation Weather (AC00-6A)
5. Aviation Weather Services (AC00-45C)
6. IFR Exam-O-Grams

A review of the information presented within this guide should provide the necessary preparation for the oral section of an FAA instrument certification or re-certification check.

1

FLIGHT PLANNING

- Certificates, Ratings and Currency Requirements
- Preflight action for flights
- Preflight action for aircraft
- Flight plan
- Route planning
- Aircraft Systems
- Fundamentals of weather
- Flight Service Stations
- Aviation weather observations and reports
- Aviation weather forecasts

A. CERTIFICATES, RATINGS AND CURRENCY REQUIREMENTS

1. AN APPLICANT FOR AN INSTRUMENT RATING MUST HAVE AT LEAST HOW MUCH AND WHAT TYPE OF FLIGHT TIME AS PILOT ? (FAR 61.65)

125 hours pilot flight time of which 50 hours are PIC cross country, 40 hours simulated or actual instrument time of which not more than 20 are in a ground trainer. 15 hours of instrument flight instruction by a CFII including 5 hours in an airplane.

2. WHEN IS AN INSTRUMENT RATING REQUIRED ? (FAR 61.3e)

When operations are conducted :

A) Under Instrument Flight Rules (IFR flight plan), or
B) In weather conditions less than the minimum for VFR flight or
C) In a positive control area or route segment.

3. WHAT ARE THE RECENCY OF EXPERIENCE REQUIREMENTS ? (FAR 61.57)

A) A Biennial flight review
B) To carry passengers, 3 takeoff's and landings within the preceding 90 days.(full stop at night)
C) 6 hours and 6 approaches within the preceding 6 months. 3 of the 6 hours must be in category. 3 hours and all approaches may be in an approved simulator.

4. IF A PILOT ALLOWS HIS/HER INSTRUMENT CURRENCY TO EXPIRE WHAT CAN BE DONE TO BECOME CURRENT AGAIN ? (FAR 61.57)

A pilot is current for the first 6 months following his/her instrument checkride or proficiency check. If the required 6 hours and 6 approaches are not accomplished within this first 6 months, you are no longer legal to file and fly under IFR. To become legal again, the FAR's allow you a "grace period" (the second 6 month period), in which you may get current by finding an "appropriately rated" safety pilot and, in simulated IFR conditions only, acquire the 6 hours and 6 approaches. If the second 6 month period also passes without doing the minimum, you may reinstate your currency by accomplishing an instrument competency check given by one of the following:

A) An FAA Inspector
B) A designated instrument flight examiner
C) A CFII

5. DEFINE "APPROPRIATELY RATED SAFETY PILOT". (AIM GLOSSARY)

This person must hold at least a Private Pilot certificate. They must also have a current medical certificate and be current in the category and class of aircraft being flown. (i.e. airplane single engine land) They need not be instrument rated.

B. PREFLIGHT ACTION FOR FLIGHT (IFR OR FLIGHT NOT IN THE VICINITY OF AIRPORT)

1. WHAT INFORMATION MUST A PILOT IN COMMAND FAMILIARIZE HIMSELF WITH BEFORE A FLIGHT ? (FAR 91.103)

All available information including :

A) Weather reports and forecasts
B) Fuel requirements
C) Alternatives if the flight cannot be completed as planned.
D) ATC delays
E) Runway lengths of intended use
F) Takeoff and landing distances

2. WHAT ARE THE FUEL REQUIREMENTS FOR FLIGHT IN IFR CONDITIONS ? (FAR 91.167)

Aircraft must carry enough fuel to fly to the first airport of intended landing (including the approach), the alternate airport (if required), and thereafter, for 45 minutes at normal cruise power. If an alternate airport is not required, enough fuel must be carried to fly to the destination airport and land with 45 minutes of fuel remaining.

C. PREFLIGHT ACTION FOR AIRCRAFT

1. WHO IS RESPONSIBLE FOR DETERMINING IF AN AIRCRAFT IS IN AIRWORTHY CONDITION ? (FAR 91.7)

The Pilot in Command

2. WHAT AIRCRAFT INSTRUMENTS/EQUIPMENT ARE REQUIRED FOR IFR OPERATIONS ? (FAR 91.205)

Those required for VFR day and night flight plus:

G enerator or alternator of adequate capacity
R adios (appropriate for facilities used)
A ltimeter (sensitive)
B all (Turn Coordinator)
C lock (sweep second hand)
A ttitude Indicator
R ate of Turn (Turn Coordinator)
D irectional Gyro

3. WHAT ARE THE REQUIRED TESTS AND INSPECTIONS OF AIRCRAFT AND EQUIPMENT TO BE LEGAL FOR IFR FLIGHT ? (FAR 91.171,91.203,91.411,91.413)

A) The aircraft must have an annual inspection. If operated for hire or rental, it must also have a 100 hour inspection. A record must be kept in the aircraft/engine logbooks.

B) The Pitot/Static system must be checked within the preceding 24 calendar months. A record must be kept in the aircraft logbook.

C) The transponder must have been checked within the preceding 24 calendar months. A record must be kept in the aircraft logbook.

D) The altimeter must have been checked within the preceding 24 calendar months. A record must be kept in the aircraft logbook.

E) The VOR must have been checked within the preceding 30 days. A record must be kept in a bound logbook.

4. MAY PORTABLE ELECTRONIC DEVICES BE OPERATED ON BOARD AN AIRCRAFT ? (FAR 91.21)

No person may operate or any PIC allow the operation of any portable electronic device:

A) On aircraft operated by an air carrier or commercial operator or
B) Any aircraft operated under IFR.

Exceptions are : portable voice recorders, hearing aids, pacemakers, electric shavers.

5. WHAT DOCUMENTS MUST BE ON BOARD AN AIRCRAFT TO MAKE IT LEGAL FOR IFR FLIGHT ? (FAR 91.203)

A irworthiness Certificate
R egistration Certificate
R adio station license
O wners manual or operating limitations
W eight & balance data

D. FLIGHT PLAN

1. WHEN IS AN IFR FLIGHT PLAN REQUIRED ? (FAR 91.135,91.173)

Anytime an aircraft is operated in controlled airspace under IFR and when operating within the positive control area or positive control area route segment.

2. WHAT ARE THE ALTERNATE AIRPORT REQUIREMENTS ? (FAR 91.169c)

1 2 3 Rule - If from 1 hour before to 1 hour after your planned ETA at the destination airport the weather is forecast to be at least 2000 ft ceilings and/or 3 mile visibilities, no alternate is required.

If less than 2000 and/or 3, an alternate must be filed using the following criteria :

A) At ETA for alternate:
For a precision approach the weather must be 600 and 2.

B) For a non-precision approach the weather must be 800 and 2.

C) For no prescribed approach the weather must allow you to descend from MEA, approach and land under basic VFR.

3. WHAT MINIMUMS ARE TO BE USED ON ARRIVAL AT THE ALTERNATE ? (FAR 91.169c)

If an instrument approach procedure has been published for that airport, the minimums specified in that procedure.

E. ROUTE PLANNING

1. WHAT ARE PREFERRED ROUTES AND WHERE CAN THEY BE FOUND ? (AIM GLOSSARY)

Routes established between busier airports to increase system efficiency and capacity. Preferred routes are listed in the Airport Facility Directory.

2. WHAT ARE ENROUTE LOW ALTITUDE CHARTS ? (AIM PARA 9-4)

Enroute Low Altitude charts provide aeronautical information for enroute instrument navigation in the low altitude route structure (up to 18000).

3. HOW OFTEN ARE ENROUTE LOW ALTITUDE CHARTS PUBLISHED ? (AIM PARA 9-4)

Every 56 days

4. WHAT ARE "AREA CHARTS"? (AIM PARA 9-4)

Area charts furnish Terminal information at a large scale in congested areas such as Dallas/Ft.Worth, Atlanta, etc. Multiple area charts are printed in one publication and are revised every 56 days.

5. WHERE CAN INFORMATION ON POSSIBLE NAVIGATIONAL AID LIMITATIONS BE FOUND? (AIM PARA 9-5)

The Airport Facility Directory.

6. WHAT OTHER USEFUL INFORMATION CAN BE FOUND IN THE AIRPORT FACILITY DIRECTORY WHICH MIGHT BE HELPFUL IN ROUTE PLANNING. (AFD)

The Airport Facility Directory contains additional information for each of the seven regions covered such as:

A) Enroute Flight Advisory Services-Locations and communications outlets.
B) ARTCC-Locations and sector frequencies.
C) Aeronautical Chart Bulletins-recent changes after publication.
D) Preferred IFR routes-high and low altitude.
E) Special notices-flight service station, GADO, Weather Service office phone numbers.
F) VOR receiver checkpoints-locations and frequencies.

7. WHAT ARE "NOTAMS" ? (AIM PARA 5-3)

Notices To Airmen - A notice containing information (not known sufficiently in advance to publicize by other means) concerning the establishment, condition, or change in any component (facility, service, or procedure of, or hazard in the National Airspace System), the timely knowledge of which is essential to personnel concerned with flight operations. They are classified into 3 categories -

A) **NOTAM (D)** - for "distant", in addition to local dissemination, are also disseminated over the entire network. They contain information on all civil public-use airports and navigational facilities that are part of the National Airspace System. Airport closures, interruption in service of navigational aids, ILS, or radar service are examples of NOTAM (D)'s.

B) **NOTAM (L)** - for "local" and describes the distribution range of the information. They are not included in hourly weather reports. NOTAM(L) information may include conditions such as taxiway closures, persons and equipment near or crossing runways, airport rotating beacon outages, and other information that would have little impact on no-local operations. Advisory or nice to know nature.

C) **FDC NOTAM** - Flight Data Center NOTAMS; Regulatory in nature; They deal with amendments to airspace, airways, instrument approach procedures, aeronautical charts and restrictions to flight. Until they are published, they are automatically included in the standard FSS briefing as are NOTAM(D)'s.

8. THE 3 BASIC TYPE OF NOTAMS CAN BE CATEGORIZED AS CLASS I OR CLASS II NOTAMS. EXPLAIN ? (AIM PARA 5-3)

Class I NOTAMS - Until a NOTAM is published in the biweekly "Notices to Airmen" publication, these NOTAMS will be considered "Class I" because they are less than two weeks old and have been distibuted solely to FSS's via the telecommunications network.
Class II NOTAMS - Once a NOTAM is published in printed form (usually occurs after two weeks as a "Class I"), it becomes a "Class II". The NOTAMS in the Class II publication are of a temporary nature but are expected to remain in effect for an extended period until expiration or cancellation.

NOTE: Published Notam's (Class II) are deleted from the data base after publication and may no longer be accessible to FSS briefers. You must take responsibility for acquiring these NOTAMS.

F. AIRCRAFT SYSTEMS

PITOT/STATIC SYSTEM

1. WHAT INSTRUMENTS OPERATE OFF OF THE PITOT/STATIC SYSTEM ? (IFH CH 4)

Altimeter, Vertical Speed, and Airspeed Indicator.

2. HOW DOES AN ALTIMETER WORK ? (IFH CH 4)

Aneroid wafers expand and contract as atmospheric pressure changes, and through a shaft and gear linkage, rotate pointers on the dial of the instrument.

3. WHAT LIMITATIONS IS A PRESSURE ALTIMETER SUBJECT TO ? (IFH CH 4)

Non-standard pressure and temperature; Temperature variations expand or contract the atmosphere and raise or lower pressure levels that the altimeter senses.

ON A WARM DAY - Pressure level is higher than standard day. Altimeter indicates lower than actual altitude.

ON A COLD DAY - The pressure level is lower than standard day. Altimeter indicates higher than actual altitude.

Changes in surface pressure also affect pressure levels at altitude.

HIGHER THAN STANDARD PRESSURE.

Pressure level is higher than standard day. Altimeter indicates lower than actual altitude.

LOWER THAN STANDARD PRESSURE.

Pressure level is lower than standard day. Altimeter indicates higher than actual altitude.

REMEMBER : HIGH TO LOW OR HOT TO COLD LOOK OUT BELOW !

4. FOR IFR FLIGHT, WHAT IS THE MAXIMUM ALLOWABLE ERROR FOR AN ALTIMETER ? (IFH CH 4)

75 Feet.

5. DEFINE AND STATE HOW YOU DETERMINE THE FOLLOWING ALTITUDES : (IFH CH 4)

Indicated altitude - Read off face of altimeter

Pressure altitude - Indicated altitude with 29.92 set in kollsman window.

True altitude - Height above sea level; Use flight computer.

Density altitude - Pressure altitude corrected for non-standard temperature Use flight computer.

Absolute altitude - Height above ground;

Subtract terrain elevation from true altitude.

6. HOW DOES THE AIRSPEED INDICATOR OPERATE ? (IFH CH 4)

It measures the difference between ram pressure from the pitot head and atmospheric pressure from the static source.

7. WHAT LIMITATIONS IS THE AIRSPEED INDICATOR SUBJECT TO ? (IFH CH 4)

Proper flow of air in the Pitot/Static system.

8. WHAT ERRORS IS THE AIRSPEED INDICATOR SUBJECT TO ? (IFH CH 4)

Position error - Caused by static ports sensing erroneous static pressure; Slipstream flow causes disturbances at static port preventing actual atmospheric pressure measurement; Varies with airspeed, altitude, configuration and may be a plus or minus value.

Density error - Changes in altitude and temperature are not compensated for by the instrument.

Compressibility error - Caused by the packing of air into the pitot tube at high airspeeds, resulting in higher than normal indications. Usually above 180 KIAS.

9. WHAT ARE THE DIFFERENT TYPES OF AIRCRAFT SPEEDS ? (IFH CH 4)

Indicated airspeed - Read off instrument

Calibrated airspeed - IAS corrected for instrument and position errors; Obtain from manual or off face of instrument.

Equivalent airspeed - CAS corrected for adiabatic compressible flow at altitude.

True airspeed - CAS corrected for non-standard temperature and pressure; Obtain from flt. computer, manual or A/S indicator slide computer.

Groundspeed - TAS corrected for wind; Speed across ground; Use flight computer.

10. ARE THE COLOR BANDS ON AN AIRSPEED INDICATOR INDICATED AIRSPEEDS OR CALIBRATED AIRSPEEDS ? (IFH CH 4)

Airspeed indicators indicate CAS if manufactured after 1976. (See owners manual to confirm).

11. HOW DOES THE VERTICAL SPEED INDICATOR WORK ? (IFH CH 4)

Changing pressures expand or contract a diaphragm connected to the indicating needle through gears and levers. The VSI is connected to the static pressure line through a calibrated leak; Measures differential pressure.

12. WHAT ARE THE LIMITATIONS OF THE VERTICAL SPEED INDICATOR ? (IFH CH 4)

It is not accurate until aircraft is stabilized. Sudden or abrupt changes in aircraft attitude will cause erroneous instrument readings as air flow fluctuates over the static port. These changes are not reflected immediately by the VSI due to calibrated leak.

13. WHAT INSTRUMENTS ARE AFFECTED WHEN THE PITOT TUBE FREEZES ? (IFH CH 4)

Airspeed Indicator only - Acts like an altimeter; It will read higher as aircraft climbs and lower as aircraft descends. It reads lower than actual in level flight.

14. WHAT INSTRUMENTS ARE AFFECTED WHEN THE STATIC PORT FREEZES ? (IFH CH 4)

Airspeed Indicator - Accurate at altitude frozen as long as static pressure in indicator and system equals outside pressure. If aircraft descends, the airspeed indicator would read high (outside static pressure greater than that trapped). If aircraft climbs, airspeed indicator reads low.

Altimeter - Indicates altitude at which system blocked.

Vertical speed - Indicates level flight

15. WHAT CORRECTIVE ACTION IS NEEDED IF PITOT TUBE FREEZES ? STATIC PORT FREEZES ? (IFH CH 4)

For pitot tube - Turn pitot heat on.

For static system - Use alternate air if available or break face of static instrument. (either VSI or A/S indicator)

16. WHAT INDICATIONS SHOULD YOU EXPECT WHILE USING ALTERNATE AIR ? (IFH CH 4)

The static pressure inside the cabin will be lower than outside pressure because of the effect of air passing the cabin at high speeds which creates a partial vacuum (venturi effect). The airspeed and altimeter will indicate higher than actual and the VSI will indicate the proper rate but in the opposite direction initially. (Pressure in aircraft lower than standard).

VACUUM/GYROSCOPIC SYSTEM

1. WHAT INSTRUMENTS OPERATE OFF OF THE VACUUM SYSTEM ? (IFH CH 4)

Normally the attitude indicator and the directional gyro. The turn coordinator could also be vacuum driven depending on the particular aircraft. The industry standard dictates that the artificial horizon and directional gyro be vacuum driven and the turn coordinator be electrically driven. However, in some systems all three can be electrically driven.

2. HOW DOES THE VACUUM SYSTEM OPERATE ? (IFH CH 4)

An engine driven vacuum pump provides suction which pulls air from the instrument case. Normal pressure entering the case is directed against rotor vanes to turn the rotor (gyro) at high speed, much like a water wheel or turbine operates. Air is drawn into the instrument through a filter from the cockpit and eventually vented outside. Vacuum values vary but provide rotor speeds from 8000 - 18000 rpm.

3. HOW DOES THE ATTITUDE INDICATOR WORK ? (IFH CH 4)

A gyro stabilizes the artificial horizon parallel to the real horizon.

4. WHAT ARE THE LIMITATIONS OF AN ATTITUDE INDICATOR ? (IFH CH 4)

The gyro will tumble above 70 degrees pitch and/or 100 degrees bank.

5. WHAT ERRORS IS THE ATTITUDE INDICATOR SUBJECT TO ? (IFH CH 4)

Errors in both pitch and bank occur during normal coordinated turns. These errors are caused by the movement of pendulous vanes by centrifugal force resulting in precession of the gyro toward the inside of the turn. The greatest error occurs in 180 degrees of turn. In a 180 degree turn to the right, on rollout the attitude indicator will indicate a slight climb and turn to the left. Acceleration and deceleration errors cause the attitude indicator to indicate a climb when the aircraft is accelerated and a descent when the aircraft is decelerated.

6. HOW DOES THE DIRECTIONAL GYRO OPERATE ? (IFH CH 4)

A gyro stabilizes the heading indicator. The speed of the gyro usually 10000 to 18000 RPM.

7. WHAT ARE THE LIMITATIONS OF THE DIRECTIONAL GYRO ? (IFH CH 4)

Beyond 55 degrees of pitch or bank the precessional force causes the card to spin rapidly.

8. WHAT ERRORS IS THE DIRECTIONAL GYRO SUBJECT TO ? (IFH CH 4)

Precession of the gyro. Maximum allowable precession is 3 degrees in 15 minutes.

ELECTRIC/GYROSCOPIC SYSTEM

1. WHAT INSTRUMENTS OPERATE ON THIS SYSTEM ? (IFH CH 4)

Turn Coordinator - This system could also operate the artificial horizon and directional gyro. Depends on the particular aircraft. (refer to "Vacuum/Gyro system" question #1).

2. HOW DOES THE TURN COORDINATOR OPERATE ? (IFH CH 4)

The turn part of the instrument uses precession to indicate direction and approximate rate of turn. A gyro reacts by trying to move in reaction to the force applied thus moving the needle or miniature aircraft in proportion to the rate of turn. The slip/skid indicator is a liquid filled tube with a ball that reacts to centrifugal force and gravity.

3. WHAT INFORMATION DOES THE TURN COORDINATOR PROVIDE? (IFH CH 4)

The miniature aircraft of the turn coordinator displays the rate of turn and rate of roll. The ball in the tube indicates a slipping or skidding condition.

Slip - Ball on the inside of turn; not enough rate of turn for amount of bank.

Skid - Ball to the outside of turn; too much rate of turn for amount of bank.

4. WHAT LIMITATIONS APPLY TO THE TURN COORDINATOR ? (IFH CH 4)

The miniature aircraft hits the stop at 45 degrees bank.

MAGNETIC COMPASS

1. HOW DOES THE MAGNETIC COMPASS WORK ? (IFH CH 4)

Magnets mounted on the compass card align themselves parallel to the earth's lines of magnetic force.

2. WHAT LIMITATIONS DOES THE MAGNETIC COMPASS HAVE ? (IFH CH 4)

18 degrees of pitch and/or bank.

3. WHAT ARE THE VARIOUS COMPASS ERRORS ? (IFH CH 4)

Oscillation error - Erratic movement of the compass card caused by turbulence or rough control technique.

Deviation error - Due to electrical and magnetic disturbances in the aircraft.

Variation error - Angular difference between true and magnetic north; Reference isogonic lines of variation.

Dip errors :

Acceleration error - On east or west headings, while accelerating, the mag. compass shows a turn to the north and when decelerating it shows a turn to the south.

REMEMBER: ANDS

A ccelerate **N** orth **D** ecelerate **S** outh

Northerly turning error - The compass leads in the south half of a turn, and lags in the north half of a turn.

REMEMBER: UNOS

U ndershoot **N** orth **O** vershoot **S** outh

G. FUNDAMENTALS OF WEATHER

1. AT WHAT RATE DOES ATMOSPHERIC PRESSURE DECREASE WITH AN INCREASE IN ALTITUDE ? (AC00-6A)

1 " Hg per 1000 ft.

2. WHAT ARE THE STANDARD TEMPERATURE AND PRESSURE VALUES FOR SEA LEVEL ? (AC00-6A)

15 degrees Centigrade and 29.92 inches mercury.

3. WITH REGARD TO THE FLOW OF AIR AROUND HIGH AND LOW PRESSURE SYSTEMS, STATE GENERAL CHARACTERISTICS IN THE NORTHERN HEMISPHERE. (AC00-6A)

Low Pressure - Inward, upward, and counter clockwise

High Pressure - Outward, downward, and clockwise

4. WHAT CAUSES THE WINDS ALOFT TO FLOW PARALLEL TO THE ISOBARS ? (AC00-6A)

The Coriolis force

5. WHY DO SURFACE WINDS GENERALLY FLOW ACROSS THE ISOBARS AT AN ANGLE ? (AC00-6A)

Surface friction

6. WHEN TEMPERATURE AND DEWPOINT ARE CLOSE TOGETHER, (WITHIN 5 DEGREES) WHAT TYPE WEATHER IS LIKELY ? (AC00-6A)

Visible moisture in the form of clouds, dew, or fog.

7. WHAT FACTOR PRIMARILY DETERMINES THE TYPE AND VERTICAL EXTENT OF CLOUDS ? (AC00-6A)

The stability of the atmosphere

8. WHAT IS THE DIFFERENCE BETWEEN A STABLE AND AN UNSTABLE ATMOSPHERE ? (AC00-6A)

An unstable atmosphere is one in which, if air is displaced vertically, will continue to move vertically, where a stable atmosphere is one which tends to resist any vertical movement of air.

9. HOW DO YOU DETERMINE STABILITY OF THE ATMOSPHERE ? (AC00-6A)

By observing the actual lapse rate and comparing to the standard lapse rate of 3.5 degrees F per 1000 ft. The "K" index of a stability chart is the primary means of determining stabilty.

10. LIST THE EFFECTS OF STABLE AND UNSTABLE AIR ON CLOUDS, TURBULENCE, PRECIPITATION AND VISIBILITY. (AC00-6A)

	STABLE	UNSTABLE
Clouds-	Stratiform	Cumiliform
Turbulence-	Smooth	Rough
Precipitation-	Steady	Showery
Visibility-	Fair to poor	Good

11. WHAT ARE THE TWO MAIN TYPES OF ICING ? (AC00-6A)

Structural and induction

12. NAME FOUR TYPES OF STRUCTURAL ICE. (AC00-6A)

Clear ice - forms when large drops strike aircraft surface and slowly freeze.

Rime ice - small drops strike aircraft and freeze rapidly.

Mixed ice - combination of above; supercooled water drops varying in size.

Frost - ice crystal deposits formed by sublimation when temperature/dewpoint is below freezing.

13. WHAT IS NECESSARY FOR STRUCTURAL ICING TO OCCUR ? (AC00-6A)

Visible moisture and below freezing temperatures.

14. WHICH TYPE OF STRUCTURAL ICING IS MORE DANGEROUS, RIME OR CLEAR ? (AC00-6A)

Clear ice; Clear ice is hard, heavy and tenacious. It is typically the most hazardous ice encountered. Clear ice forms when, after initial impact, the remaining liquid portion of the drop flows out over the aircraft surface gradually freezing as a smooth sheet of solid ice. This type forms when drops are large as in rain or in cumuliform clouds. It's removal by de-icing equipment is especially difficult due to the fact it forms as it flows away from the de-icing equipment (inflatable boots etc.).

15. WHAT FACTORS MUST BE PRESENT FOR A THUNDERSTORM TO FORM ? (AC00-6A)

A) A source of lift (heating, fast moving front)
B) Unstable air (non-standard lapse rate)
C) High moisture content (temp/dewpoint close)

16. WHAT ARE "SQUALL LINE" THUNDERSTORMS ? (AC00-6A)

A "squall line" is a non-frontal, narrow band of active thunderstorms. Often it develops ahead of a cold front in moist, unstable air, but it may develop in unstable air far removed from any front. The line may be too long to easily detour and too wide and severe to penetrate. It often contains severe steady-state thunderstorms and presents the single most intense weather hazard to aircraft. It usually forms rapidly, generally reaching a maximum intensity during the late afternoon and the first few hours of darkness.

17. STATE TWO BASIC WAYS FOG MAY FORM. (AC00-6A)

A) Cool air to dewpoint
B) Adding moisture to air

18. NAME SEVERAL TYPES OF FOG. (AC00-6A)

A) Radiation fog
B) Advection fog
C) Upslope fog

D) Steam fog
E) Precipitation-Induced fog

19. WHAT CAUSES RADIATION FOG TO FORM ? (AC00-6A)

The ground cools adjacent air to dewpoint on calm, clear nights.

20. WHAT IS ADVECTION FOG AND WHERE IS IT MOST LIKELY TO FORM ? (AC00-6A)

Advection fog results from the transport of warm humid air over a cold surface. A pilot can expect advection fog to form primarily along coastal areas during the winter. Unlike radiation fog, it may occur with winds, cloudy skies, over a wide geographic area, and at anytime of day or night.

21. DEFINE UPSLOPE FOG ? (AC00-6A)

Upslope fog forms when air flows upward over rising terrain and is, consequently, adiabatically cooled to or below its initial dewpoint. It is commonly found along the western slopes of the Rocky mountains.

22. DEFINE STEAM FOG ? (AC00-6A)

Steam fog occurs as a result of the movement of cold air over warm water. A good example of this type of fog may be found over and around the Great Lakes area. If the air above is very cold, that air will tend to be unstable, with heating from below causing the air to rise. If the air is considerably unstable (abnormal lapse rate), thunderstorms and turbulence may be expected. If only the lower layers are unstable, a dense low cloud mass is most likely with some icing potential.

23. WHAT IS PRECIPITATION-INDUCED FOG ? (AC00-6A)

Precipitation-induced fog is the result of evaporation of rain or drizzle, evaporation occurring while precipitation is falling and/or it has reached the ground. This type fog is prevalent especially when associated with warm fronts, although it may occur with or without the presence of fronts.

24. WHY IS FOG A MAJOR OPERATIONAL CONCERN TO PILOTS ? (AC00-6A)

It is of primary concern during takeoffs and landings. Fog can reduce vertical and horizontal visibilities to as little as zero-zero. It can occur instantly from a clear condition, making takeoffs, landings, and even taxiing, potentially hazardous operations.

H. FLIGHT SERVICE STATIONS

1. WHAT ARE THE STANDARD FLIGHT SERVICE STATION FREQUENCIES ? (AIM PARA 4-44)

121.5 and 122.2 are standard FSS freqs. 122.0 (Flight Watch) may be available at FSS designated. 123.6 may be available at fields at which FSS is located (Airport Advisory Area). (123.6 may not be available at fields with an operative control tower.)

2. HOW LONG BEFORE DEPARTURE SHOULD YOU FILE AN IFR FLIGHT PLAN ? (AIM PARA 5-7)

Pilots should file IFR flight plans at least 30 minutes prior to the estimated departure time.

3. WHAT IS AN AIRPORT ADVISORY AREA ? (AIM PARA 3-62)

An area within 10 nm of an airport without a tower or the tower is not in operation and on which a FSS is located. The FSS will provide advisories on active runway, altimeter setting, NOTAMS, etc.

4. WHAT IS "FLIGHT WATCH" ? (AIM GLOSSARY)

Also known as Enroute Flight Advisory Service (EFAS), it is designed to provide, on pilot request, weather information pertinent to the type of flight intended, route and altitude flown. The frequency is 122.0.

I. AVIATION WEATHER OBSERVATIONS AND REPORTS

1. IN SURFACE AVIATION WEATHER REPORTS. (AWS 2)

VISIBILITIES ARE : STATUTE OR NAUTICAL?

CLOUD HEIGHTS ARE : AGL OR MSL?

WIND DIRECTIONS ARE : TRUE OR MAGNETIC NORTH?

WIND SPEEDS ARE : KNOTS OR MILES PER HOUR?

Statute, AGL, True North*, Knots.

* Wind direction for the local station is broadcast in degrees magnetic. i.e. ATIS, Tower.

2. HOW OFTEN ARE SURFACE OBSERVATIONS ISSUED ? (AWS 2)

A) Record observations (SA) on the hour.
B) Special reports (RS,SP), observations taken on the hour or when needed to report significant changes in weather.

3. WHAT ARE PIREP'S AND WHERE ARE THEY USUALLY FOUND ? (AWS 3)

Abbreviation for "Pilot Report". Contains information concerning cloud tops, icing, and turbulence as observed by pilots enroute. A PIREP is usually transmitted as part of a group of PIREPS collected by state or as a remark appended to a surface aviation weather report.

4. THE CLOUD BASES AND TOPS IN "PIREPS" ARE EXPRESSED IN MSL OR AGL ? (AWS 3)

Above sea level; The pilot reads MSL altitudes from the altimeter when making the report.

5. DEFINE THE FOLLOWING TYPES OF VISIBILITIES.

FLIGHT VISIBILITY - The average forward horizontal distance, from the cockpit of an aircraft in flight, at which prominent unlighted objects may be seen and identified by day and prominent lighted objects may be seen and identified by night.

PREVAILING VISIBILITY - defined as the greatest horizontal visibility equaled or exceeded throughout at least half the horizon circle which need not necessarily be continuous. Normally found in hourly aviation weather reports. This value may be considerably different than the actual slant range visibility the pilot would see when approaching a runway due to local obscuring phenomena such as fog or low clouds.

RUNWAY VISIBILITY RANGE (RVR) - defined as the maximum horizontal distance down a specified runway at which a pilot can see and identify standard high intensity runway lights. It is always determined using a transmissometer and is reported in hundreds of feet. It is reported in the remarks section of hourly sequence reports whenever the prevailing visibility is less than 2 miles, or the RVR is 6,000 ft. or less.

RUNWAY VISIBILITY VALUE (RVV) - defined as the visibility determined for a particular runway by a transmissometer. A meter provides a continuous indication of the visibility (reported in miles or fractions of miles) for the runway. RVV is used in lieu of prevailing visibility in determining minimums for a particular runway. It is also reported in the remarks of hourly sequence reports when visibilities become less than 2 miles.

6. GIVE SOME EXAMPLES OF CURRENT WEATHER CHARTS AVAILABLE AT THE FSS OR NWSO USED IN IFR FLIGHT PLANNING. (AWS 5 THRU 15)

- Surface Analysis Chart
- Weather Depiction Chart
- Freezing Level Chart
- Stability Chart
- Radar Summary Chart

7. WHAT IS A SURFACE ANALYSIS CHART ? (AWS 5)

The Surface Analysis Chart provides a ready means of locating pressure systems and fronts and also gives an overview of winds and temperature/dewpoint spreads at the time of observation.

8. WHAT INFORMATION DOES A WEATHER DEPICTION CHART PROVIDE ? (AWS 6)

This chart contains only part of surface information. It depicts cloud coverage, ceilings and visibilities at a glance. It is best used as a general look at areas of low ceilings and visibilities and for determining where the nearest VFR is located.

9. WHY ARE RADAR SUMMARY CHARTS IMPORTANT IN PILOT WEATHER BRIEFINGS ? (AWS 7)

The Radar Summary chart deals primarily with weather of a potentially hazardous nature. Anything shown on this chart along or near a pilot's route of flight must be taken into consideration and considered carefully. Radar observations are taken hourly. They are transmitted 17 times in each 24 hours. The charts show actual areas of radar echoes which are produced by a concentration of liquid or frozen water drops. These echoes represent the interior regions of moisture-laden clouds and, the greater the concentration and size of the drops (as in cumulonimbus clouds), the stronger the echoes and the greater the probability of hazards.

10. WHAT INFORMATION IS PROVIDED BY A FREEZING LEVEL CHART ? (AWS 10)

A Freezing Level chart is an analysis of observed freezing level data from upper air observations. Solid lines on the chart depict contours of the freezing level drawn for 4,000 foot intervals and labelled in hundreds of feet MSL. The contour analysis shows an overall view of the lowest observed freezing level. Always plan for possible icing in clouds or precipitation above the freezing level — especially between temperatures of 0 degrees Celsius and -10 degrees Celsius. Area forecasts show more specifically the areas of expected icing. Low level significant weather progs show anticipated changes in the freezing level.

11. WHAT IS A STABILITY CHART USEFUL FOR ? (AWS 11)

The stability chart outlines areas of stable and unstable air. Two stability indices are computed for each upper air station; one is the "lifted index" and the other, the "K index". From the chart you can make a quick estimate of areas of probable convective turbulence as well as areas of clouds and precipitation.

J. AVIATION WEATHER FORECASTS

1. WHAT ARE TERMINAL FORECASTS (FT) ? (AWS 4)

A terminal forecast (FT) is a description of the surface weather expected to occur at an airport. The forecast cloud heights and amounts, visibility, weather and wind relate to flight operations within 5 nautical miles of the center of the runway complex. Scheduled forecasts are issued by the Weather Service Forecast Office for their respective areas 3 times daily and are valid for 24 hours.

2. WHAT ARE AREA FORECASTS (FA) ? (AWS 4)

An area forecast (FA) is a forecast of general weather conditions over an area the size of several states.It is used to determine forecast enroute weather and to interpolate conditions at airports which do not have FT's issued. FA's are issued 3 times a day by the National Aviation Weather Advisory unit in Kansas City. Each FA consists of a 12 hour forecast plus a 6 hour outlook. All times are Greenwich Mean Time (GMT). All distances except visibility are in nautical miles. Visibility is in statute miles. The FA is comprised of 5 sections -

A) HAZARDS/FLIGHT PRECAUTIONS (H)
B) SYNOPSIS (S)
C) ICING (I)
D) TURBULENCE (T)
E) LOW LEVEL WIND SHEAR (IF APPLICABLE)
F) SIGNIFICANT CLOUDS AND WEATHER (C)

3. WHAT ARE 12 AND 24 HOUR PROGNOSTIC CHARTS ? (AWS 8)

Significant weather prognostic charts, portray forecast weather which may influence flight planning. Significant weather progs. are prepared for the conterminous U.S. and adjacent areas. The U.S. low level significant weather prog. is designed for domestic flight planning up to 24,000 ft. The low level prog. is a 4 panel chart. The 2 lower panels are 12 & 24 hour surface progs. The 2 upper panels are 12 & 24 hour progs. of significant weather from the surface to 400 millibars (24,000 ft.). The charts show conditions as they are forecast to be at the valid time of the chart.

4. DEFINE THE TERMS: LIFR, IFR, MVFR AND VFR. (AWS 4)

LIFR = Ceilings less than 500 and/or visibilities less than 1 mile.

IFR = Ceilings less than 1000 and/or visibilities less than 3 miles.

MVFR = Ceilings less than 3000 and/or visibilities less than 5 miles.

VFR = Ceilings greater than 3000 and visibilities greater than 5 miles.

5. WHAT VALUABLE INFORMATION CAN BE DETERMINED FROM WINDS AND TEMPERATURES ALOFT FORECASTS ? (AWS 4)

A) Can determine most favorable altitude based on winds and direction of flight.
B) Can determine areas of possible icing by noting air temperatures of +2 degrees centigrade to - 20 degrees centigrade.
C) Determine temperature inversions.
D) Determine turbulence by observing abrupt changes in wind direction and speed at different altitudes.

6. WHAT IS A SIGMET ? (AWS 4)

Sigmet's advise of weather potentially hazardous to all aircraft.

A) Severe icing
B) Severe or extreme turbulence

C) Widespread sand or duststorms lowering visibilities to less than 3 miles.

7. WHAT IS AN AIRMET ? (AWS 4)

An Airmet is for weather that may be hazardous to single engine and light aircraft.

A) Moderate icing.
B) Moderate turbulence.
C) Sustained winds at the surface of 30 knots or more.
D) Ceilings less than 1000 ft and/or visibility less than 3 miles affecting over 50% of the area at one time.
E) Extensive mountain obscurement.

8. WHAT IS A CONVECTIVE SIGMET ? (AWS 4)

Convective Sigmet's imply severe or greater turbulence, severe icing and low level wind shear.

A) Tornadoes.
B) Lines of thunderstorms.
C) Embedded thunderstorms.
D) Large hail.

9. WHAT IS "PATWAS" ? (AIM PARA 7-7)

Pilots Automatic Telephone Weather Answering Service (PATWAS). PATWAS is provided by nonautomated flight service stations. PATWAS. PATWAS is a continuous recording of meteorological and aeronautical information which is available at selected locations by telephone. The recording contains a summary of data for an area within 50 NM of the parent station.

10. WHAT IS "TIBS" ? (AIM PARA 7-7)

Telephone Informaton Briefing Service (TIBS). TIBS is provided by automated flight service stations and provides continuous telephone recordings of meteorological and/or aeronautical information. Specifically, TIBS provides area and/or route briefings, airspace procedures, and special announcements concerning aviation interests.

2

DEPARTURE

- Authority and limitations of pilot
- IFR flight plan
- Departure clearance
- Departure procedures
- VOR accuracy checks
- Transponder
- Airport facilities

A. AUTHORITY AND LIMITATIONS OF PILOT

1. DISCUSS FAR 91.3 "RESPONSIBILITY AND AUTHORITY OF PIC". (FAR 91.3)

The Pilot in Command of an aircraft is directly responsibility for and is final authority as to, the operation of that aircraft.

2. WHAT ARE THE RIGHT OF WAY RULES PERTAINING TO IFR FLIGHTS ? (FAR 91.113)

When weather conditions permit, regardless of whether an operation is under IFR or VFR, vigilance shall be maintained by each person operating an aircraft so as to see and avoid other aircraft.

3. WHAT ARE THE REQUIRED REPORTS FOR EQUIPMENT MALFUNCTION UNDER IFR IN CONTROLLED AIRSPACE ? (FAR 91.187)

A) Loss of VOR,TACAN,ADF, or Low Freq. Nav. receiver capability.
B) Complete or partial loss of ILS receiver capability.
C) Impairment of air/ground communication capability.
D) Loss of DME above 24,000 feet.

B. IFR FLIGHT PLAN

1. WHEN MUST A PILOT FILE AN IFR FLIGHT PLAN ? (AIM PARA 5-7)

Prior to departure from within or prior to entering controlled airspace a pilot must submit a complete flight plan and receive clearance from ATC if weather conditions are below VFR minimums. The pilot should file at least 30 minutes prior to departure.

2. WHEN CAN YOU CANCEL YOUR IFR FLIGHT PLAN ? (AIM PARA 5-13)

In controlled airspace, anytime you are operating in VFR conditions, and when operations are being conducted outside of a Positive Control area.

3. WHAT IS A COMPOSITE FLIGHT PLAN ? (AIM PARA 5-6)

Flight plans which specify VFR operation for one portion of a flight, and IFR for the other.

4. WHAT 4 PIECES OF EQUIPMENT DETERMINE YOUR "SPECIAL EQUIPMENT" SUFFIX WHEN FILING AN IFR FLIGHT PLAN ? (AIM PARA 5-7)

Transponder, Mode C or altitude encoding capability, DME, and RNAV.

C. DEPARTURE CLEARANCE

1. HOW CAN YOUR IFR CLEARANCE BE OBTAINED ? (AIM PARA 5-7)

Airports with an ATC tower in operation :

Clearance may be received from either ground control or a specific clearance delivery frequency when available.

Airports without a tower or FSS on the field, or in an outlying area :

There are several methods of obtaining IFR clearances at non-tower, non-FSS, and outlying airports.

A) Clearances may be received over the radio thru a RCO (remote comm. outlet) or, in some cases, over the telephone.

B) In some areas, a clearance delivery frequency is available that is usable at different airports within a particular geographic area, such as a TCA for example.

C) If the above methods are not available, your clearance can be obtained, once airborne, from ARTCC provided you remain VFR in controlled airspace.

The procedure may vary due to geographical features, weather conditions, and the complexity of the ATC system. To determine the most effective means of receiving an IFR clearance, pilots should ask the nearest FSS the most appropriate means of obtaining their IFR clearance.

2. WHAT DOES "CLEARED AS FILED" MEAN ? (AIM PARA 5-23)

ATC will issue abbreviated IFR clearance based on the route of flight as filed in IFR flight plan, provided the filed route can be approved with little or no revision.

3. WHICH CLEARANCE ITEMS ARE GIVEN IN AN ABBREVIATED IFR CLEARANCE ? (AIM PARA 5-23)

C learance Limit (destination airport or fix)
R oute (initial heading)
A ltitude (initial altitude)
D eparture Freq.
S quawk (transponder code)

4. WHAT DOES "CLEARANCE VOID TIME" MEAN ? (AIM PARA 5-24)

When operating from an airport without a tower, a pilot may receive a clearance containing a provision that if the flight has not departed by a specific time the clearance is void.

D. DEPARTURE PROCEDURES

1. WHAT IS A "SID" ? (AIM PARA 5-26)

Standard Instrument Departure Procedure - An IFR departure procedure printed for pilot use in graphic or textual form. It provides a transition procedure from the terminal to enroute structure.

2. MUST YOU ACCEPT A "SID" IF ASSIGNED ONE ? (AIM PARA 5-26)

No; If you do not possess a textual description or graphic depiction of the SID you cannot accept one. Also, if for any other reason you cannot accept the SID, you must advise ATC. ATC prefers you let them know ahead of time in the "remarks" section of your flight plan.

3. WHAT MINIMUMS ARE NECESSARY FOR IFR TAKEOFF UNDER FAR 91 REGS. AND UNDER FAR 121,129,135 REGS. ? (FAR 91.175)

Except when non-standard takeoff minimums apply as noted on the approach chart, none for FAR Part 91.

Under FAR 121,129 and 135 :

For 2 engines or less .... 1 SM visibility
For more than 2 engines .... 1/2 SM visibility

4. HOW DOES A PILOT DETERMINE IF NON-STANDARD TAKEOFF MINIMUMS EXIST AT A PARTICULAR AIRPORT ? (AIR PARA 5-26)

A large "T" in a black triangle at the bottom of the approach chart of the departure airport indicates non-standard takeoff minimums or departures exist.

5. WHAT IS CONSIDERED "GOOD OPERATING PRACTICE" IN DECIDING ON TAKEOFF MINIMUMS FOR IFR FLIGHT ?

If an instrument approach procedure has been prescribed for that airport, use the minimums for that approach for takeoff. If no approach procedure is available, basic VFR minimums would be recommended. (1000 and 3)

6. WHERE CAN A PILOT FIND NON-STANDARD TAKEOFF MINIMUMS IF THEY APPLY ? (AIM PARA 5-26)

On US Government approach charts, non-standard takeoff minimums may be found in the front section of the book. On Jeppesen approach charts, non-standard takeoff minimums are on the particular approach chart being used.

7. WHAT IS THE STANDARD MINIMUM CLIMB GRADIENT FOR ALL STANDARD INSTRUMENT DEPARTURES ? (AIM PARA 5-26)

200 ft. per nautical mile.

8. WHEN A SID SPECIFIES A CLIMB GRADIENT IN EXCESS OF 200 FT PER NAUTICAL MILE, WHAT SIGNIFICANCE SHOULD THIS HAVE TO THE PILOT ? (AIM PARA 5-26)

Climb gradients are specified when required for obstacle clearance. Crossing restrictions in the SID's may be established for traffic separation or obstacle clearance. When no gradient is specified the pilot is expected to climb at least 200 ft. per nautical mile to the MEA unless required to level off by a crossing restriction.

9. A CLIMB GRADIENT OF 300 FT. PER NAUTICAL MILE AT A GROUNDSPEED OF 100 KNOTS REQUIRES WHAT RATE OF CLIMB ? (AIM SID CHART)

498 ft per minute

GS / 60 * Climb Gradient = Feet per Minute

100 divided by 60 = 1.66 times 300 = 498 FPM

E. VOR ACCURACY CHECKS

1. WHAT ARE THE DIFFERENT METHODS FOR CHECKING THE ACCURACY OF VOR EQUIPMENT ? (FAR 91.171 NEW)

A) VOT Check; Plus or minus 4 degrees
B) Ground checkpoint; Plus or minus 4 degrees
C) Airborne checkpoint; Plus or minus 6 degrees
D) Dual VOR check; 4 degrees within each other
E) Select a radial over a known ground point; Plus or minus 6 degrees
F) Repair station can use radiated test signal, but only technician performing test can make entry in logbook.

2. WHAT RECORDS MUST BE KEPT CONCERNING VOR CHECKS ? (FAR 91.171)

Each person making a VOR check shall enter the date, place, and bearing error and sign the aircraft log or other reliable record.

3. WHERE CAN A PILOT FIND THE LOCATION OF THE NEAREST VOT TESTING STATION ? (AIM PARA 1-4)

An Airport/Facility Directory or the tab on a Enroute Low Altitude Chart.

F. TRANSPONDER

1. WHERE IS A TRANSPONDER (MODE C) REQUIRED ? (FAR 91.215)

A) All aircraft operations in all airspace at and above 10,000ft MSL except in that airspace at and below 2,500ft AGL.

B) All aircraft operations within the airspace designated as a Terminal Control Area (TCA).

C) All aircraft operations within a 30 mile radius of the primary airport for which a TCA is designated, from the surface to 10,000ft MSL or the upper limits of the TCA, whichever is higher.

D) All aircraft operations within and above an ARSA up to and including 10,000ft MSL.

E) All aircraft operations within a 10 mile radius of certain airports, from the surface to 10,000ft MSL, excluding that airspace below 1,200ft AGL beyond the Airport Traffic Area.

2. WHAT ARE THE FOLLOWING TRANSPONDER CODES ? (AIM PARA 6-12)

1200 .... VFR
7700 .... Emergency
7600 .... Communications Emergency
7500 .... Hijacking in progress

3. WHAT TRANSPONDER PROCEDURE SHOULD BE FOLLOWED IN THE EVENT OF A TWO WAY COMMUNICATIONS FAILURE ? (AIM PARA 6-31)

Squawk 7700 for 1 minute; Then squawk 7600 for 15 minutes; Repeat steps 1 & 2 as long as necessary.

G. AIRPORT FACILITIES

1. WHERE CAN A PILOT FIND INFORMATION CONCERNING FACILITIES AVAILABLE FOR A PARTICULAR AIRPORT ? (AIM PARA 9-5)

Airport Facilities Directory; Contains information concerning services available, communication data, navigational facilities, special notices etc.

2. WHAT ARE THE FOLLOWING ABBREVIATIONS ? (AIM PARA 2-1 & 2-2)

ALS........	Approach Light System
VASI	Visual Approach Slope Indicator
PAPI	Precision Approach Path Indicator
REIL	Runway End Identifier Lights

3. WHAT COLOR ARE RUNWAY EDGE LIGHTS ? (AIM PARA 2-4)

White; On instrument runways amber replaces white on the last 2000 feet or half the runway length, whichever is less.

4. WHAT COLORS AND COLOR COMBINATIONS ARE STANDARD AIRPORT ROTATING BEACONS ? (AIM PARA 2-8)

Lighted Land Airport	White/Green
Lighted Water Airport	White/Yellow
Military Airport	2 White/Green

5. THE OPERATION OF A ROTATING BEACON AT AN AIRPORT WITHIN A CONTROL ZONE DURING THE HOURS OF DAYLIGHT MEANS WHAT ? (AIM PARA 2-8)

Ground visibility is less than 3 miles and/or the ceiling is less than 1000 feet; ATC clearance is required to conduct flight operations within.

6. WHERE WOULD INFORMATION CONCERNING RUNWAY LENGTHS, WIDTHS, AND WEIGHT BEARING CAPACITIES, BE FOUND ? (AFD)

Airport/Facility Directory.

7. THE TOUCHDOWN ZONE MARKING ON A PRECISION INSTRUMENT RUNWAY IS HOW FAR FROM THE THRESHOLD? (AIM PARA 2-22)

Touchdown zone markings are a pair of 3 parallel stripes on either side of the runway centerline and are located 500ft from the landing threshold. Normally, the standard glide slope angle of 3 degrees, if flown to the surface, will assure touchdown within this zone.

8. WHAT IS THE PURPOSE OF FIXED DISTANCE MARKINGS? (AIM PARA 2-21)

Fixed distance markings provide a visual aiming point. The pilot can estimate a visual glidepath that will intersect the marking, assuring a landing within the 3000ft touchdown zone.

9. THE TOUCHDOWN ZONE EXTENDS HOW FAR DOWN A RUNWAY? (AIM PARA 2-21)

The touchdown zone extends 3000ft past the threshold. The touchdown zone elevation (TDZE) is the highest point in the zone.

3

ENROUTE

- Enroute limitations
- Enroute procedures
- Oxygen requirements
- Emergencies
- Radio orientation
- Attitude instrument flying
- Unusual flight conditions
- Radio navigation
- Airway route system
- Special use airspace
- Enroute weather services
- Physiological factors

A. ENROUTE LIMITATIONS

1.DEFINE THE FOLLOWING : (AIM GLOSSARY)

MEA - Minimum Enroute Altitude; Lowest published altitude between radio fixes which assures acceptable navigational signal coverage and meets obstacle clearance requirements.

MOCA - Minimum Obstacle Clearance Altitude - Lowest published altitude between radio fixes that meets obstacle clearance requirements and which assures acceptable navigational signal coverage within 22 NM of VOR.

MCA - Minimum Crossing Altitude; Lowest altitude at certain fixes at which aircraft must cross when preceding in direction of a higher MEA.

MRA - Minimum Reception Altitude; Lowest altitude at which an intersection can be determined.

MAA - Maximum Authorized Altitude; Maximum altitude useable for a route segment that assures signal reception without interference from another signal on same freq.

2. IF NO APPLICABLE MINIMUM ALTITUDE IS PRESCRIBED (NO MEA OR MOCA), WHAT MINIMUM ALTITUDES APPLY FOR IFR OPERATIONS ? (FAR 91.177)

Mountainous terrain - At least 2000 feet above the highest obstacle within 5 SM of course to be flown. Other than mountainous terrain - At least 1000 feet above the highest obstacle within 5 SM of course to be flown.

3.WHAT CRUISING ALTITUDES SHALL BE MAINTAINED WHILE OPERATING UNDER IFR IN CONTROLLED AIRSPACE ? UN-CONTROLLED AIRSPACE ? (FAR 91.179)

IFR flights within controlled airspace shall maintain the altitude or flight level assigned by ATC. In un-controlled airspace altitude is selected based on magnetic course flown.

Below 18000 MSL:

0 - 179 degrees ..... Odd thousand MSL
180 - 359 degrees ... Even thousand MSL

18000 up to 29000 MSL:

0 - 179 degrees ..... Odd Flight levels
180 - 359 degrees ... Even Flight levels

4.WHAT ALTITUDES APPLY WHEN OPERATING ON AN IFR FLIGHT PLAN AND ATC ASSIGNS "VFR ON TOP" ? (FAR 91.159,91.179)

The pilot shall maintain the appropriate VFR altitudes unless otherwise assigned by ATC.

5.WHAT PROCEDURES ARE APPLICABLE CONCERNING COURSES TO BE FLOWN WHEN OPERATING IFR ? (FAR 91.181)

Except when maneuvering an aircraft to pass well clear of other aircraft or the maneuvering of an aircraft in VFR conditions to clear intended flight path, the following applies :

A) Maintain centerline of Federal airway
B) Maintain direct route between navigational aids or fixes defining route.

6.WHAT IS THE POSITIVE CONTROL AREA ? WHAT ARE THE REQUIREMENTS TO OPERATE WITHIN ? (FAR 91.135)

The Positive Control Area is airspace in which there is positive control of all aircraft. It extends from 18000 MSL to FL 600 and not less than 1500 AGL. To operate within:

A) Must be on an IFR flight plan
B) IFR equipped aircraft
C) Instrument rated pilot

D) Mode C Transponder and 2 way radio

B. ENROUTE PROCEDURES

1. WHAT REPORTS SHOULD BE MADE TO ATC AT ALL TIMES WITHOUT A SPECIFIC REQUEST? (AIM PARA 5-33)

A) True airspeed changes 5 % or 10 knots, which ever is greater.
B) Unable to climb or descend at 500 fpm
C) Loss of NAV, ILS, or radio comm. capability
D) Safety of flight, anything affecting
E) Altitude, leaving assigned
F) Holding, leaving or entering
G) Altitude, changes VFR on top
H) Missed Approach

2. WHAT REPORTING REQUIREMENTS ARE REQUIRED BY ATC IN A NON-RADAR ENVIRONMENT ? (AIM PARA 5-33)

In addition to the above, the following :

A) Final Approach Fix inbound
B) Time estimate in excess of 3 minute error

3. IF UN-FORECAST WEATHER IS ENCOUNTERED ENROUTE ARE YOU REQUIRED TO REPORT ? (AIM PARA 5-33)

Yes

4. THE CLEARANCE TO "CRUISE 9000" HAS WHAT MEANING ? (AIM PARA 4-83)

Used in an ATC clearance to authorize a pilot to conduct flight at any altitude from minimum IFR altitude up to and including 9000 ft. The pilot may level off at any intermediate altitude. Climb and descent within the block is at the discretion of the pilot. However, once the pilot starts a descent and VERBALLY reports leaving an altitude, he may not return to that altitude without ATC clearance. This clearance can also be an approval to proceed to and make an approach at the destination airport.

5. WHEN OPERATING "VFR ON TOP" PILOTS ON IFR FLIGHT PLANS MUST ADHERE TO WHAT OPERATIONAL PROCEDURES ? (AIM PARA 4-87)

A) Fly an appropriate VFR altitude
B) Comply with VFR visibility requirements
C) Comply with IFR regs. (Min. IFR altitudes, position reports, etc.)

6. WHAT IS A "CLEARANCE LIMIT" AND WHEN DO YOU RECEIVE ONE ? (AIM PARA 4-83)

A "Clearance Limit" is a fix, point, or location to which an aircraft is cleared when issued an ATC clearance. You will normally get cleared to a fix other than the destination airport when ATC is experiencing delays.

7. WHAT INFORMATION WILL BE PROVIDED BY ATC WHEN REQUESTED TO HOLD AT A FIX WHERE THE HOLDING PATTERN IS NOT CHARTED ? (AIM PARA 5-37)

A) Direction of hold from fix
B) Holding fix name
C) Radial course or bearing to hold on
D) Leg length in miles if DME or RNAV equipped
E) Direction of turns (if non-standard)
F) Expect further clearance time (EFC)

8. WHAT IS THE MAXIMUM SPEED FOR AN AIRCRAFT WHILE HOLDING ? (AIM PARA 5-37)

For propeller driven aircraft ... 175 knots

9. WHAT IS THE MEANING OF NON-STANDARD VS. STANDARD HOLDING PATTERNS ? (AIM PARA 5-37)

Standard .... All turns to the right
Non-standard .... All turns to the left

10. DESCRIBE THE PROCEDURE FOR CROSS WIND CORRECTION IN A HOLDING PATTERN ? (AIM PARA 5-37)

The pilot compensates for wind effect primarily by drift correction on the inbound and outbound legs. On the outbound leg, double the inbound correction to avoid major turning adjustments.

11. WHAT ACTION IS APPROPRIATE WHEN APPROACHING A HOLDING FIX AT AN AIRSPEED IN EXCESS OF MAXIMUM HOLDING SPEED ? (AIM PARA 5-37)

Start speed reduction when 3 minutes or less from fix. Speed may be reduced earlier, but ATC must be advised of the change.

12. WHY IS IT IMPORTANT FOR THE PILOT TO RECEIVE AN EFC TIME WITH INITIAL HOLDING INSTRUCTIONS ? (AIM PARA 5-37)

In case of lost communications, you will know when to depart hold.

13. DESCRIBE THE DIFFERENT ENTRY METHODS FOR HOLDING ? (AIM PARA 5-37) (FIG 4-11 AIM)

Parallel
Teardrop
Direct

14. AT 14,000 FT AND BELOW, LEG LENGTH FOR A STANDARD HOLDING PATTERN IS WHAT ? (AIM PARA 5-37)

1 minute inbound below 14,000
1 1/2 minutes inbound above 14,000

15. IF ASSIGNED A DME HOLD, WHAT PROCEDURES SHOULD BE UTILIZED ? (IFH CH 12)

The same entry and holding procedures apply to DME holding except distances (nautical miles) are used instead of time values. The outbound course of the DME holding pattern is called the outbound leg of the pattern. The length of the outbound leg will be specified by the controller and the end of this leg is determined by the DME readout.

C. OXYGEN REQUIREMENTS

1. WHAT REGULATIONS APPLY CONCERNING SUPPLEMENTAL OXYGEN ? (FAR 91.211)

At cabin pressure altitudes above 12,500 MSL up to and including 14000 MSL, the crew after 30 minutes.

Above 14000 MSL up to and including 15000 MSL, the minimum flight crew must be continuously on oxygen.

Above 15000 MSL, each passenger must be provided with supplemental oxygen and the minimum flight crew must be continuously on oxygen.

D. EMERGENCIES

1. WHEN MAY THE PILOT IN COMMAND OF AN AIRCRAFT DEVIATE FROM AN ATC CLEARANCE ? (FAR 91.123)

Except in an emergency, no person may, in an area which ATC is exercised, operate an aircraft contrary to an ATC instruction.

2. IF AN EMERGENCY ACTION REQUIRES DEVIATION FROM AN FAR 91 REG., MUST A PILOT SUBMIT A WRITTEN REPORT, AND IF SO, TO WHOM ? (FAR 91.123)

Each Pilot in Command who is given priority by ATC in an emergency shall, if requested by ATC, submit a detailed report of that emergency within 48 hours, to the chief of that ATC facility.

3. CONCERNING TWO WAY RADIO COMMUNICATIONS FAILURE IN VFR AND IFR CONDITIONS, WHAT IS THE PROCEDURE FOR ALTITUDE, ROUTE, LEAVING HOLDING FIX, DESCENT FOR APPROACH, AND APPROACH SELECTION ? (FAR 91.185)

In VFR conditions :

If the failure occurs in VFR, or if VFR is encountered after the failure, each pilot shall continue the flight under VFR and land as soon as practicable.

In IFR conditions :

Altitude - Whichever altitude is highest for each segment of the flight being flown.

A) Altitude or FL last assigned
B) The MEA
C) The altitude or FL ATC has advised to expect

Route - By route assigned in the last ATC clearance.

A) If being radar vectored, by a direct route from point of radio failure to the fix, route or airway specified in the vector clearance.

B) In absence of an assigned route, a route that ATC has advised may be expected.

C) The route filed in the flight plan.

Leaving holding fix/clearance limit -

A) If an EFC time has been received, leave holding fix/clearance limit at that time and proceed to and hold, if necessary, at the holding pattern depicted for the IAP, or if none is depicted, at the fix from which the approach begins.

B) If no EFC is received, the pilot is expected to proceed to and hold, if necessary, at the holding pattern depicted for the IAP or, if none is depicted, at the IAF from which the approach begins. If more than one IAF is available, it is the pilot's choice and ATC protects all of them.

Descent for the approach -

A) Begin descent from the enroute altitude or flight level upon reaching the IAF, but not before the ETA shown on the flight plan as amended with ATC.

B) If holding is necessary, at the radio fix to be used for the approach at the destination airport, holding and descent to the initial approach altitude is done in accordance with the pattern depicted on the approach chart. If none is depicted, holding and descent is done in a holding pattern on the side of the final approach course on which the procedure turn is described.

Selection of approach -

A) That which ATC has advised may be expected.
B) In absence of above, any approach the pilot chooses.

4. ASSUMING TWO-WAY COMMUNICATIONS FAILURE, DISCUSS THE RECOMMENDED PROCEDURE TO FOLLOW CONCERNING ALTITUDES TO BE FLOWN FOR THE FOLLOWING TRIP.

The MEA between A and B is 5,000 ft. The MEA between B and C is 5,000 ft. The MEA between C and D is 11,000 ft. The MEA between D and E is 7,000 ft. You have been cleared via A,B,C,D, to E. While flying between A and B, your assigned altitude was 6,000 ft. and you were told to expect a clearance to 8,000 ft. at B. Prior to receiving the higher altitude assignment, you experience two-way communication failure.

The correct procedure would be as follows :

A) Maintain 6,000 ft. to B, then climb to 8,000 ft. (the altitude you were advised to expect.)

B) Continue to maintain 8,000 ft. then climb to 11,000 ft. at C, or prior to C if necessary to comply with an MCA at C.

C) Upon reaching D, you would descend to 8,000 ft. (even though the MEA was 7,000 ft.), as 8,000 ft. was the highest of the altitude situations stated in the rule.

5. WHAT PROCEDURE WOULD YOU USE IF ALL COMMUNICATION AND NAVIGATION EQUIPMENT FAILED (COMPLETE ELECTRICAL SYSTEM FAILURE) ?

A) First determine you have complete loss. Determine cause (check circuit breakers, alternator, ammeter, etc.)

B) Review the preflight weather briefing for the nearest VFR; Determine heading and altitude and proceed to VFR conditions, using VFR altitudes.

C) If VFR conditions are not within range of aircraft, get off the airway and determine heading to an unpopulated area relatively free of obstructions (terrain or man-made, i.e. rural areas, large lakes, ocean, etc.)

D) Establish decent on a specific heading to VFR conditions; proceed VFR to nearest airport.

E. RADIO ORIENTATION

1. WHAT ANGULAR DEVIATION FROM A VOR COURSE IS REPRESENTED BY 1/2 SCALE DEFLECTION OF THE CDI ? (IFH CH 8)

5 degrees; Full scale deflection = 10 degrees

2. WHAT DISTANCE OFF COURSE WOULD AN AIRCRAFT BE WITH 1/2 SCALE DEFLECTION 30 MILES OUT ? (IFH CH 8)

Aircraft displacement from course is approximately 200 feet per dot per mile.

Example: $\dfrac{200 \times (2.5 \times 30)}{6000}$ $\dfrac{15000}{6000} = 2.5$ NM off

Remember: 1 dot 30 miles out = 1 NM off.
1 dot 60 miles out = 2 NM off.

3.HOW DO YOU DETERMINE TIME AND DISTANCE FROM A VOR/NDB STATION? (IFH CH8)

A) Determine the radial on which you are located.
B) Turn 80 degrees right, or left, of the inbound course, rotating the OBS to the nearest 10 degree increment opposite the direction of turn.
C) Maintain heading. When the CDI centers, note the time.
D) Maintaining the same heading, rotate the OBS 10 degrees in the same directions as above.
E) Note the elapsed time when the CDI again centers.
F) Time/distance from the station is determined from the following formulas:

Time to station:

$$\frac{\text{TIME IN SECONDS BETWEEN BEARING CHG}}{\text{DEGREES IN BEARING CHG}}$$

Distance to station:

$$\frac{\text{TAS TIMES MINUTES FLOWN}}{\text{DEGREES OF BEARING CHG}}$$

4.WHAT DEGREE OF ACCURACY CAN BE EXPECTED IN VOR NAVIGATION ? (IFH CH 8)

Plus or Minus 1 Degree.

5.HOW DO YOU FIND AN ADF RELATIVE BEARING ? (IFH CH 8)

A Relative Bearing is the angular relationship between the aircraft heading and the station measured clockwise from the nose; The bearing is read directly on the ADF dial measured clockwise from zero.

6.HOW DO YOU FIND AN ADF MAGNETIC BEARING ? (IFH CH 8)

A Magnetic Bearing is the direction of an imaginary line from the aircraft to the station or the station to the aircraft referenced to magnetic north. To determine use this formula :

MH + RB = MB

(Mag. heading + Rel. bearing = Mag. bearing)

If the sum is more than 360, subtract 360 to get the magnetic bearing To the station. The reciprocal of this number is the magnetic bearing from the station.

7.WHAT IS ADF HOMING ? (IFH CH 8)

Flying the aircraft on any heading required to keep the ADF needle on zero until station is reached.

8.WHAT IS ADF TRACKING ? (IFH CH 8)

A procedure used to fly a straight geographic flight path inbound to or outbound from an NDB. A heading is established that will maintain the desired track.

F. ATTITUDE INSTRUMENT FLYING

1.WHAT ARE THE THREE FUNDAMENTAL SKILLS INVOLVED IN ATTITUDE INSTRUMENT FLYING ? (IFH CH 5)

- Cross check instruments (scan)
- Instrument interpretation
- Aircraft control

2.ATTITUDE INSTRUMENT FLYING INVOLVES GROUPING INSTRUMENTS AS THEY RELATE TO CONTROL FUNCTION AS WELL AS AIRCRAFT PERFORMANCE. WHAT ARE THE THREE MAJOR GROUPS AND WHAT INSTRUMENTS ARE IN EACH ? (IFH CH 5)

PITCH	BANK	POWER
Attitude Ind.	Attitude Ind.	Man. Press. Guage
Altimeter	Dir. Gyro	Tachometer
Airspeed	Turn Coord.	Airspeed
Vert. Speed		

3.WHAT INSTRUMENTS ARE PRIMARY FOR PITCH, BANK, AND POWER IN STRAIGHT AND LEVEL FLIGHT ? (IFH CH 5)

Altimeter, Directional Gyro, and Airspeed

4.DESCRIBE THE PROCEDURE FOR LEVELING OFF FROM A DESCENT AT A LOW AIRSPEED AND A HIGH AIRSPEED (IFH CH 5)

Low airspeed descent: Begin adding power 150 ft above target altitude.
High airspeed descent : Begin level off 50 feet above target altitude.

G. UNUSUAL FLIGHT CONDITIONS

1.IF A THUNDERSTORM WERE ENCOUNTERED INADVERTENTLY, WHAT FLIGHT INSTRUMENT SHOULD BE USED TO MAINTAIN CONTROL OF AIRCRAFT ? WHAT PROCEDURE ? (AW CH 11)

Attitude Indicator; Establish power for the recommended maneuvering speed and attempt to maintain a constant attitude only. Do not attempt to maintain a constant altitude.

2.WHAT ARE THE CONDITIONS NEEDED FOR MAJOR STRUCTURAL ICING TO FORM ? (AW CH 10)

- Visible moisture
- Air temperature near or below freezing
- Freezing aircraft surface

3.WHAT ACTION IS RECOMENDED IF YOU INADVERTENTLY ENCOUTER ICING CONDITIONS? (AC00-6A)

- Change course and/or altitude.
- Usually climb to a higher altitude.

4.WHICH TYPE PRECIPITATION WILL PRODUCE THE MOST HAZARDOUS ICING CONDITIONS ? (AW CH 10)

Freezing rain.

H. RADIO NAVIGATION

1. WHAT FREQUENCY RANGE DO VOR'S OPERATE WITHIN ? (IFH CH 7)

VHF 108.0 to 117.95 MHz.

2. WHAT ARE THE NORMAL USABLE DISTANCES FOR THE VARIOUS CLASS VOR STATIONS ? (IFH CH 7)

Omniranges are classified according to their operational uses. The standard VOR facility has a power output of approximately 200 watts, with a maximum usable range depending upon the aircraft altitude, class of facility, location and siting of the facility, terrain conditions within the usable area of the facility and other factors. Above and beyond certain altitude and distance limits, signal interference from other VOR facilities and signal weakening make the signal unreliable.

H-VOR's and L-VOR's have a normal usable distance of 40 nautical miles below 18,000 ft. T-VOR's are short range facilities which have a power output of approximately 50 watts and a usable distance of 25 nautical miles at 12,000 ft. and below. T-VOR's are used primarily in terminal areas, on or adjacent to airports, for instrument approaches.

Terminal class	= 1000 to 12000 AGL	25 NM
Low class	= 1000 to 18000 AGL	40 NM
High class	= 1000 to 14500 AGL	40 NM
High class	= 14500 to 18000 AGL ...	100 NM
High class	= 18000 to 45000 AGL ...	130 NM
High class	= 45000 to 60000 AGL ...	100 NM

3. WHAT IS THE MEANING OF A SINGLE CODED IDENTIFICATION RECEIVED ONLY ONCE EVERY 30 SECONDS FROM A VORTAC STATION ? (AIM PARA 1-7,1-14)

The DME component is operative; The VOR component inoperative. It is important to recognize which identifier is retained for the operative facility; A single coded identifier with a repeat interval every 30 seconds, indicates DME is operative. If no identification is received, the facility has been taken off the air for tune-up or repair even though intermittent or constant signals are received.

4.WILL ALL VOR STATIONS HAVE CAPABILITY FOR PROVIDING DISTANCE INFORMATION TO AIRCRAFT EQUIPPED WITH DME ? (AIM PARA 1-7)

No; Aircraft receiving equipment assures reception of azimuth and distance information from a common source only when designated as VOR/DME, VORTAC, ILS/DME, AND LOC/DME stations.

5.FOR OPERATIONS OFF ESTABLISHED AIRWAYS AT 17,000 FEET MSL, H CLASS VORTAC FACILITIES USED TO DEFINE A DIRECT ROUTE OF FLIGHT SHOULD BE NO FURTHER THAN WHAT DISTANCE FROM EACH OTHER ? (AIM PARA 5-7)

200 NM.

6.WHAT FREQUENCY RANGE DO NDB'S NORMALLY OPERATE WITHIN ? (AIM PARA 1-2)

Low to medium Freq. .... 190 to 535 KHz.

7.WHEN A RADIO BEACON IS USED IN CONJUNCTION WITH AN ILS MARKER BEACON, WHAT IS IT CALLED ? (AIM PARA 1-2)

A Compass Locator.

8.THERE ARE FOUR TYPES OF NDB FACILITIES IN USE. WHAT ARE THEY AND WHAT ARE THERE EFFECTIVE RANGES ? (IFH CH 8)

HH facilities 2000 watts .... 75 NM
H facilities 50 to 1999 watts .... 50 NM
MH facilities Less than 50 watts .... 25 NM
ILS compass locator Less than 25 watts .. 15 NM

9.WHAT LIMITATIONS APPLY WHEN USING AN NDB FOR NAVIGATION ? (AIM PARA 1-2)

Radio beacons are subject to disturbances from lightning, precipitation static etc. At night radio beacons are vulnerable to interference from distant stations.

10. WHAT OPERATIONAL PROCEDURE WOULD BE UTILIZED WHEN NAVIGATION OR APPROACHES ARE CONDUCTED USING AN NDB ? (AIM PARA 1-2)

Since ADF receivers do not incorporate signal flags to warn a pilot when erroneous bearing information is being displayed, the pilot should continuously monitor the NDB's code identification.

11. WHAT IS "DME" ? (AIM PARA 1-7)

Distance Measuring Equipment; Aircraft equipped with DME are provided with distance and groundspeed information when receiving a VORTAC or TACAN facility. In the operation of DME paired pulses at a specific spacing are sent out from the aircraft and are received at the ground station. The ground station then transmits paired pulses back to the aircraft at the same pulse spacing but on a different frequency. The time required for the round trip of this signal exchange is measured in the airborne DME unit and is translated into distance and groundspeed. Reliable signals may be received at distances up to 199 NM at line of sight altitude. DME operates on frequencies in the UHF spectrum between 962 MHz and 1213 MHz. Distance information is slant range distance not horizontal.

12. WHAT IS "RNAV" ? (AIM GLOSSARY)

RNAV is an abbreviation for Area Navigation and is a form of navigation that permits aircraft properly equipped, to operate on any desired course within the coverage of station-referenced navigation signals. This form of navigation allows a pilot to select a more direct course to a destination by not requiring overflight of ground based navigational aids. Navigation is to selected "way points" instead of VOR's. A "waypoint" is created by simply moving the VOR to a point along route of flight desired.

13. WHAT IS "LORAN" ? (AIM PARA 1-17)

Long Range Navigation - A navigational system based upon measurement of the difference in time of arrival of pulses of radio signals radiated by a group of transmitter stations which are separated by hundreds of miles. A LORAN receiver is basically an on-board computer capable of determining the aircraft's position based on the measurement of time-differences receipt of these different signals. LORAN receivers also have computer memory capable of storing information and useful programs such as:

A) Airport locations
B) Location of airspace such as TCA's etc.
C) Navigational aids (VOR's, NDB's)
D) Navigational fixes (intersections,waypoints)
E) Ability to continuously compute nearest airport, or bearing to for emergency use.
F) Distance from starting point and destination
G) Time enroute and estimated time to station.
H) Groundspeed and true airspeed

I. AIRWAY ROUTE SYSTEM

1.THE VICTOR AIRWAY SYSTEM CONSISTS OF AIRWAYS EXTENDING FROM WHAT ALTITUDES ? (IFH CH 10)

1200 AGL up to but not including 18000 MSL.

2.HOW WIDE IS A VICTOR AIRWAY ? (AIM PARA 5-34)

8 NM.

3.WHAT IS THE DIFFERENCE BETWEEN EVEN AND ODD NUMBERED AIRWAYS ? (IFH CH 10)

Like highways, Victor Airways are designated by number-Generally North/South airways are odd; East/West airways are even. When airways coincide on the same radial, the airway segment shows the numbers of all the airways on it.

4.WHAT IS THE SIGNIFICANCE OF A LETTER FOLLOWING THE AIRWAY NUMBER ? (IFH CH 10)

It designates that airway as an alternate airway and where it is in relation to the main victor airway.

5.WHAT IS A "CHANGEOVER POINT" ? (AIM PARA 5-36)

It is a point along the route or airway segment between 2 adjacent navigational facilities or waypoints where changeover in navigational guidance should occur.

6. WHAT IS A "WAYPOINT" ? (AIM PARA 5-34)

It designates a RNAV fix.

7. ARE THE COURSES DEPICTED ON A LOW ALTITUDE ENROUTE CHART MAGNETIC OR TRUE COURSES ? (IFH CH 13))

Magnetic.

8. DESCRIBE THE CLIMB PROCEDURE WHEN APPROACHING A FIX, BEYOND WHICH, A HIGHER MEA EXISTS. (IFR Exam-O-Gram)

A pilot may begin a climb to the new MEA at, or after, reaching the fix.

9. DESCRIBE THE CLIMB PROCEDURE WHEN APPROACHING A FIX AT WHICH A MCA EXISTS. (IFR Exam-O-Gram)

A pilot should begin a climb when approaching the fix, so as to arrive at that fix at the MCA. A Minimum Crossing Altitude is the minimum altitude at which certain radio facilities or intersections must be crossed in specified directions of flight. If a normal climb, commenced immediately after passing a fix beyond which a higher MEA applies, would not assure adequate obstruction clearance, an MCA is specified.

10. FOR THE FOLLOWING TERMS, IDENTIFY THE SYMBOLS WHICH CORRESPOND TO THEM ON ENROUTE LOW ALTITUDE CHARTS.

TACAN

VOR

VORTAC

ATC noncompulsory reporting point

ATC compulsory reporting point

Compass rose

MOCA

MEA

DME fix distance when not obvious

DME fix distance when same as route miles

Mileage break at airway course change

VOR changeover point

Mileage between intersections, VOR's, breakdown points

Mileage between VOR's or a VOR and compulsory reporting point

Victor airway

Remote air/ground communications with ARTCC

Special use airspace

Airport without published instrument approach

Localizer facility information box

Compass locator frequency

Commercial broadcast station

VORTAC facility information box

Special VFR (fixed wing) in control zone not authorized

Airport with published instrument approach

ARTCC boundary

Change in MEA, MOCA at other than nav. aids

Controlling ARTCC

Minimum crossing altitude

Minimum reception altitude

NDB

Maximum authorized altitude

Magnetic variation

Control zone boundary

Controlling FSS

J. SPECIAL USE AIRSPACE

1. DEFINE THE FOLLOWING TYPES OF AIRSPACE ? (AIM PARA 3-42 thru 3-46)

Prohibited Area - Aircraft flight prohibited

Restricted Area - Must have permission from controlling authority if VFR flight. IFR flights will be cleared thru or vectored around.

Military Operations Area - Permission not required but VFR flights should exercise caution. IFR flights will be cleared thru or vectored around.

Warning Area - Established beyond 3 mile limit International Airspace; Permission not required but flight plan advised. Same hazards as restricted areas.

Alert Area - Airspace containing a high volume of pilot training or unusual aerial activity. No permission required but VFR flights should exercise caution. IFR flights will be cleared thru or vectored around.

K. ENROUTE WEATHER SERVICES

1.EFAS WHAT IS IT ? (AIM PARA 7-4)

Also known as Flight Watch, Enroute Flight Advisory Service provides, upon request, timely weather information pertinent to type of flight, intended route of flight and altitude. Frequency is 122.00.

2."TWEB", WHAT IS IT ? (AIM PARA 7-8)

Transcribed Weather Broadcast . Continuous recording of meteorological and aeronautical information broadcast on L/MF and VOR facilities.

3.WHERE CAN A PILOT FIND FREQUENCIES / AVAILABILITY OF EFAS AND TWEB ? (AIM PARA 7-4)

Airport Facilities Directory.

L. PHYSIOLOGICAL FACTORS

1.WHAT IS HYPOXIA ? (AIM PARA 8-2)

Hypoxia is a state of oxygen deficiency in the body sufficient to impair functions of the brain and other organs.

2.WHAT FACTORS CAN MAKE YOU MORE SUCCEPTIBLE TO HYPOXIA ? (AIM PARA 8-2)

Alcohol, low doses of certain drugs such as antihistamines, tranquilizers, sedatives, analgesics; also cigarette smoking.

3.FOR OPTIMUM PROTECTION AGAINST HYPOXIA, PILOTS SHOULD USE SUPPLEMENTAL OXYGEN WHEN ? (AIM PARA 8-2)

During Daylight hours .... Above 10000 ft
At Night .... Above 5000 ft

4. WHAT IS HYPERVENTILATION ? (AIM PARA 8-2)

An abnormal increase in the volume of air breathed in and out of the lungs. Causes an excessive loss of carbon dioxide from body. Symptoms are lightheadedness, drowsiness, tingling in extremities.

5. WHAT IS RECOMMENDED IF HYPERVENTILATION IS SUSPECTED ? (AIM PARA 8-2)

The rate and depth of breathing should be consciously slowed down. Controlled breathing in and out of a paper bag can build carbon dioxide back to a normal level.

6. WHAT IS CARBON MONOXIDE POISONING ? (AIM PARA 8-4)

Carbon monoxide is a colorless, odorless, and tasteless gas contained in exhaust fumes. Most heaters in light aircraft work by air flowing over the exhaust manifold in which cracks may form causing carbon monoxide to leak into cockpit.

7. WHAT ARE SYMPTOMS OF CARBON MONOXIDE POISONING AND WHAT ACTION SHOULD BE TAKEN IF SUSPECTED ? (AIM PARA 8-4)

Headache, drowsiness, dizziness
Shut heater off; Open air vents.

4

ARRIVAL

- Approach control
- Precision approaches
- Non-precision approaches
- Circling approaches
- Missed approaches
- Logging flight time
- Instrument approach procedure chart - general
- Instrument approach procedure chart - planview
- Instrument approach procedure chart - profile
- Instrument approach prodedure chart - minimums section
- Instrument approach procedure chart - aerodrome

A. APPROACH CONTROL

1. WHAT IS A STAR ? (AIM PARA 5-41)

Standard Terminal Arrival Route - A STAR is an ATC coded IFR arrival route established for application to arriving IFR aircraft destined for certain airports. It's purpose is to simplify clearance delivery procedures.

2. IF ATC ISSUES YOUR FLIGHT A STAR, MUST YOU ACCEPT IT ? (AIM PARA 5-41)

No; If you do decide to accept, you must have an approved textual description available. Recommended action if STAR is not desired: Place "NO STAR" in Remarks section of IFR flight plan.

3. WHEN BEING RADAR VECTORED FOR AN APPROACH, AT WHAT POINT MAY YOU START A DESCENT FROM YOUR LAST ASSIGNED ALTITUDE TO A LOWER ALTITUDE IF "CLEARED FOR THE APPROACH" ? (FAR 91.116i)

When established on a segment of a published route or Instrument Approach Procedure.

4. DEFINE THE TERMS : (AIM GLOSSARY)
INITIAL APPROACH SEGMENT
INTERMEDIATE APPROACH SEGMENT
FINAL APPROACH SEGMENT
MISSED APPROACH SEGMENT

An instrument approach procedure may have as many as four separate segments depending on how the approach procedure is structured.

A) **Initial Approach Segment** : That segment of an instrument approach procedure between the initial approach fix and the intermediate approach fix or, where applicable, the final approach fix or point.

B) **Intermediate Approach Segment** : That segment of an instrument approach procedure between either the intermediate approach fix and the final approach fix or point, or between the end of a reversal, race track or dead reckoning track procedure and the final approach fix or point, as appropriate.

C) **Final Approach Segment** : That segment of an instrument approach procedure in which alignment and descent for landing are accomplished.

D) **Missed Approach Segment** : The segment between the missed approach point or the point of arrival at decision height and the missed approach fix at the prescribed altitude.

5. WHAT OBSTACLE CLEARANCE ARE YOU NORMALLY GUARANTEED DURING THE INITIAL, INTERMEDIATE, FINAL, AND MISSED APPROACH SEGMENTS OF A STANDARD INSTRUMENT APPROACH PROCEDURE ?

The Initial Approach Segment provides 1000 ft. obstacle clearance when within 4 miles either side of course.

The Intermediate Approach Segment provides 500 ft. obstacle clearance within a protected area tapering uniformly from the initial approach segment to the width of the final approach course.

The Final Approach Segment obstacle clearance depends on the type of approach. ILS approaches have the least with 190 ft. being the standard. NDB approaches require the most obstacle clearance usually being as much as 350 ft. The width of the obstacle clearance area depends on the type of approach as well as the length of the final approach course.

6. HOW DOES A PILOT NAVIGATE BETWEEN THE ENROUTE PHASE AND THE INITIAL APPROACH SEGMENT ? (AIM PARA 5-46,5-47)

Feeder Routes or Radar Vectors. Feeder routes are depicted on approach procedure charts to designate routes for aircraft to proceed from the enroute structure to the initial approach fix.

7. EXPLAIN THE TERMS "MAINTAIN" VS. "CRUISE" AS THEY PERTAIN TO AN IFR ALTITUDE ASSIGNMENT ? (AIM PARA 4-83)

Maintain - Self-explanatory; Maintain last altitude assigned.

Cruise - Used instead of "maintain" to assign a block of airspace to a pilot, from minimum IFR altitude up to and including the altitude specified in the cruise clearance. The pilot may level off at any intermediate altitude, and climb/descent may be made at the discretion of the pilot. However, once the pilot starts a descent, and VERBALLY reports leaving an altitude in the block, he may not return to that altitude without additional ATC clearance.

8. WHAT PROCEDURE IS TO BE USED WHEN THE CLEARANCE "CLEARED FOR THE VISUAL" IS ISSUED ? (AIM PARA 5-58)

It allows a pilot to discontinue the instrument approach procedure for a particular runway, and proceed to the airport visually, provided VFR conditions exist (1000 and 3), and the pilot has the airport in sight or the aircraft in front of him in sight. It may be initiated by ATC without request.

9. DESCRIBE THE TERM "CONTACT APPROACH" ? (AIM PARA 5-60)

With ATC authorization, allows a pilot to deviate from an instrument approach procedure and proceed to the airport visually.

It requires :

- 1 NM flight visibility.
- Remain clear of clouds.
- Must maintain visual with ground.
- It must be requested by the pilot; ATC will not initiate a "contact approach".

10. WHEN IS A PROCEDURE TURN NOT REQUIRED ? (AIM PARA 5-48)

- When radar vectored inbound.
- From an authorized holding pattern.
- When the procedure specifies NoPT.
- When a teardrop PT is depicted.
- As restricted by notes and symbols on chart.

11. WHAT ARE STANDARD PROCEDURE TURN LIMITATIONS ? (AIM PARA 5-48)

- Must turn on the depicted side.
- Follow depicted altitudes.
- Remain within 10 NM of navigational facility utilizes.

B. PRECISION APPROACHES

1. DEFINE A PRECISION APPROACH ? (AIM GLOSSARY)

A standard Instrument Approach Procedure in which an electronic glideslope/glidepath is provided. Examples are : ILS, MLS, and PAR.

2. WHAT ARE THE BASIC COMPONENTS OF A STANDARD ILS ? (AIM PARA 1-10)

Guidance .... Localizer, Glide Slope
Range .... Marker beacons
Visual ...... Approach lights, Touchdown
Centerline lights, Runway edge lights

3. DESCRIBE BOTH VISUAL AND AURAL INDICATIONS A PILOT WOULD RECEIVE WHEN CROSSING THE OUTER, MIDDLE, AND INNER MARKERS OF A STANDARD ILS. (AIM PARA 1-10)

Outer Marker	**Middle Marker**	**Inner Marker**
Blue light	Amber light	White light
Dull tone	Medium tone	High tone
Slow speed	Medium speed	High speed
- - - - - -	- . - . - .	. . . . . .

4. WHAT ARE THE DISTANCES FROM THE LANDING THRESHOLD OF THE OUTER, MIDDLE, AND INNER MARKERS ? (AIM PARA 1-10)

Outer marker ...... 4 to 7 miles from threshold.
Middle marker ..... 3500 ft. from threshold.
Inner marker ...... Between MM and threshold.

5. WHEN IS THE INNER MARKER USED ? (AIM PARA 1-10)

When a Category II approach is conducted; On at DH of 100 ft. on the glide slope.

6. TO MAINTAIN GLIDE SLOPE AND DESIRED AIRSPEED ON AN ILS APPROACH, HOW ARE POWER AND PITCH USED ? (IFH CH 5)

Change pitch to control altitude.
Change power to control airspeed.

7. WHAT IS THE FREQUENCY RANGE OF A LOCALIZER ? (AIM PARA 1-10)

108.10 to 111.95 MHz (odd tenths).

8. THE LOCALIZER / TRANSMITTER ANTENNA INSTALLATION IS LOCATED WHERE IN RELATION TO THE RUNWAY ? (AIM PARA 1-10)

The antenna is located at the far end of the approach runway.

9. THE GLIDE SLOPE TRANSMITTER ANTENNA IS LOCATED WHERE IN RELATION TO THE RUNWAY ? (AIM PARA 1-10)

The glide slope transmitter is located between 750 ft and 1250 ft from the approach end of the runway (down the runway), and offset 250 ft to 650 ft from it.

10. WHAT RANGE DOES A STANDARD LOCALIZER HAVE ? (IFH CH 7)

The localizer signal provides course guidance throughout the descent path to the runway threshold from a distance of 18 NM from the antenna site.

11. WHAT IS THE ANGULAR WIDTH OF A LOCALIZER SIGNAL ? (IFH CH 7)

The localizer signal is adjusted to provide an angular width of between 3 and 6 degrees as necessary, to provide a linear width of 700 ft at the runway approach threshold.

12. WHAT IS THE NORMAL GLIDE SLOPE ANGLE FOR A STANDARD ILS ? (IFH CH 7)

3 degrees and a depth of 1.4 degrees.

13. WHAT IS THE SENSITIVITY OF A CDI TUNED TO A LOCALIZER SIGNAL COMPARED TO A CDI TUNED TO A VOR ? (IFH CH 7)

Full left or full right deflection occurs at approximately 2 1/2 degrees from the centerline of a localizer course. CDI sensitivity is 4 times greater than when tuned to a VOR where full scale deflection equals 10 degrees from the centerline.

14. DEFINE THE TERM "DECISION HEIGHT". (FAR PART 1)

With respect to the operation of an aircraft, it means the height at which a decision must be made, during an ILS, MLS, or PAR instrument approach, to either continue the approach or execute a missed approach.

15. WHEN FLYING AN INSTRUMENT APPROACH PROCEDURE, WHEN CAN THE PILOT DESCEND BELOW THE MDA OR DH ? (FAR 91.175)

No person may operate an aircraft below the prescribed MDA or continue an approach below the authorized DH unless :

A) Aircraft is continuously in a position from which a descent to a landing on intended runway can be made at a normal rate of descent using normal maneuvers.

B) The flight visibility is not less than the visibility prescribed in the standard instrument approach procedure being used.

C) When at least one of the following visual references for the intended runway is distinctly visible and identifiable to pilot :

1) The approach light system, except that the pilot may not descend below 100 feet above the touchdown zone

elevation using the ALS as a reference unless the red terminating bars or the red side row bars are also distinctly visible and identifiable.

2) Threshold.
3) Threshold markings.
4) Threshold lights.
5) REIL.
6) VASI.
7) Touchdown zone markings.
8) Touchdown zone lights.
9) Runway and runway markings.
10) Runway lights.

16. WHAT ARE THE LEGAL SUBSTITUTES FOR AN ILS OUTER MARKER AND MIDDLE MARKER ? (FAR 91.175)

Outer marker: Compass locator, PAR, ASR or DME, VOR and NDB fixes authorized in the instrument approach procedure.

Middle marker: Compass locators or PAR are the only legal substitutions.

17. WHAT ARE PAR AND ASR APPROACHES ? (AIM PARA 5-50)

A PAR approach is a type of radar approach in which a controller provides highly accurate navigational guidance in azimuth and elevation to the pilot. (Precision approach) An ASR approach is a type of radar approach in which a controller provides navigational guidance in azimuth only. (Non-precision approach)

18. WHAT IS A "NO GYRO" APPROACH ? (AIM GLOSSARY)

A radar approach/vector provided in case of a malfunctioning gyro-compass or DG; Instead of providing the pilot with headings to be flown, the controller observes radar track and issues control instructions "Turn Right/Left" or "Stop Turn" as appropriate.

19. WHAT RATE OF TURN IS RECOMMENDED DURING EXECUTION OF A "NO GYRO" APPROACH PROCEDURE ? (AIM PARA 5-50)

Standard rate until on final; Half standard rate on final approach.

20. IF CONDUCTING AN ASR APPROACH ARE THE MINIMUMS EXPRESSED AS DH OR MDA ? (AIM PARA 5-50)

An ASR approach is a Non-precision approach with no glide slope provided; Minimums are depicted as MDA.

C. NON-PRECISION APPROACHES

1. WHAT IS A NON-PRECISION APPROACH ? (AIM GLOSSARY)

A standard instrument approach procedure in which no glide slope is provided.

2. NAME THE TYPES OF NON-PRECISION APPROACH PROCEDURES AVAILABLE ? (AIM GLOSSARY)

VOR, TACAN, NDB, LOC, ASR, LDA, and SDF.

3. DEFINE "MDA". (FAR PART 1)

Minimum Descent Altitude; The lowest altitude, expressed in feet above MSL, to which descent is authorized on final approach where no electronic glide slope is provided, or during circle to land maneuvering. In execution of a standard instrument approach procedure, MDA figures are rounded to 20 feet increments MSL.

4. DEFINE "VDP" . (AIM GLOSSARY)

Visual Descent Point; A "VDP" is a defined point on the final approach course of a non-precision straight-in approach procedure from which normal descent from the MDA to the runway touchdown point may be commenced, provided the approach threshold of that runway, or approach lights or other markings identifiable with the approach end of that runway are clearly visible to pilot.

5. **WILL STANDARD INSTRUMENT APPROACH PROCEDURES ALWAYS HAVE A FAF ? (IFH AC 90-1A EXERPT)**

No; NDB and VOR approaches with the primary navigational aid on the field will not always have a designated FAF.

6. **IF NO FAF IS PUBLISHED, WHERE DOES THE FINAL APPROACH SEGMENT BEGIN ON A NON-PRECISION APPROACH ? (IFH AC90-1A EXERPT)**

Where the procedure turn intersects the final approach course inbound.

7. **WHEN CAN A PILOT MAKE A STRAIGHT IN LANDING IF EXECUTING AN IAP HAVING ONLY CIRCLING MINIMUMS ? (AIM PARA 5-56)**

A straight-in landing may be made if the pilot has the runway in sight in sufficient time to make a normal approach for landing and has been cleared to land.

8. **A STRAIGHT IN LANDING PROCEDURE NORMALLY REQUIRES THE FINAL APPROACH COURSE AND RUNWAY ALIGNMENT TO BE WITHIN HOW MANY DEGREES ? (AIM PARA 5-56)**

30 degrees.

9. **WHAT IS A STEPDOWN FIX ? (AIM GLOSSARY)**

A fix permitting additional descent within a segment of an instrument approach procedure by identifying a point at which a controlling obstacle has been safely overflown.

10. **WHAT DOES A VASI SYSTEM PROVIDE ? (AIM PARA 2-2)**

Provides visual descent guidance during an approach to a runway.
Provides safe obstacle clearance within plus or minus 10 degrees of extended runway centerline and to 4 NM from the runway.
Provides a 3 degree slope.

11. WHAT ARE THE MAJOR DIFFERENCES BETWEEN SDF AND LDA APPROACHES ? (IFH CH 7)

SDF Approach Procedure : The SDF course may or may not be aligned with runway; Usable off-course indications are limited to 35 degrees either side of course centerline; The SDF signal emitted is fixed at either 6 or 12 degrees.

LDA Approach Procedure : The LDA course is of comparable utility and accuracy to a standard localizer; An LDA course is usually not aligned with runway, however straight-in minima may be published where the angle between the centerline and course does not exceed 30 degrees; If the angle exceeds 30 degrees, circling minimums are published.

12. WHAT CRITERIA DETERMINES WHETHER OR NOT YOU MAY ATTEMPT AN APPROACH ? (FAR 91.175)

No regulation states that you cannot attempt an approach, if operating under Part 91 regulations. But if you reach MDA or DH and decide to descend below to land, flight visibility must be at least equal to that published.

13. WHAT FAR'S REQUIRE USE OF SPECIFIED PROCEDURES BY ALL PILOTS APPROACHING FOR LANDING UNDER IFR ? (FAR)

FAR PART 97.

D. CIRCLING APPROACHES

1. WHAT ARE CIRCLE TO LAND APPROACHES ? (AIM GLOSSARY)

Not technically an approach but a maneuver initiated by a pilot to align aircraft with runway for landing when a straight-in landing from an instrument approach is not possible or desirable; The maneuver is made only when authorized by ATC and/or visual reference with airport is established.

2. WHEN DO CIRCLING MINIMUMS APPLY ? (AIM PARA 5-56)

When either the normal rate of descent from MDA would be excessive or the runway alignment factor of 30 degrees is exceeded, a straight-in minimum is not published and a circling minimum applies.

3. IF CLEARED FOR A "STRAIGHT IN VOR-DME 34 APPROACH" CAN A PILOT CIRCLE TO LAND IF NEEDED ? (AIM GLOSSARY)

A "Straight-In Approach" is an instrument approach wherein final approach is begun without first having executed a procedure turn, not necessarly completed with a straight-in landing or made to straight-in minimums.

4. WHAT ARE THE RESTRICTIONS APPLICABLE TO A CIRCLE TO LAND MANEUVER ? (AIM PARA 5-56)

- Must maintain visual with airport at all times.
- Remain at or above circling minimums until aircraft is continuously in a position to land.

5. WHILE CIRCLING TO LAND YOU LOSE VISUAL WITH THE RUNWAY ENVIRONMENT. YOUR APPROXIMATE POSITION AT THE TIME VISUAL CONTACT IS LOST, IS A BASE LEG AT THE CIRCLING MDA. WHAT PROCEDURE SHOULD BE FOLLOWED ? (AIM PARA 5-57)

If visual reference is lost while circling to land from an instrument approach, the pilot should make an initial climbing turn toward the landing runway and continue the turn until established on the missed approach course. Inasmuch as the circling maneuver may be accomplished in more than one direction, different patterns will be required to become established on the prescribed missed approach course, depending on the aircraft position at the time visual reference is lost. Adherence to the procedure will assure that an aircraft will remain within the circling and missed approach obstacle clearance areas.

6. WHAT OBSTACLE CLEARANCE ARE YOU GUARANTEED DURING A CIRCLING APPROACH MANEUVER ?

In all circling approaches, the circling minimum provides 300 ft. of obstacle clearance within the circling approach area. The size of this area depends on the category in which the aircraft operates within.

Category A	1.3 mile radius
Category B	1.5 mile radius
Category C	1.7 mile radius
Category D	2.3 mile radius
Category E	4.5 mile radius

E. MISSED APPROACHES

1. WHEN MUST A PILOT EXECUTE A MISSED APPROACH ? (AIM PARA 5-75)

Execution of a missed approach must occur when one of the following conditions exist :

A) Arrival at missed approach point and runway environment not in sight.
B) Arrival at DH on glideslope and runway environment not in sight.
C) Anytime you determine a safe landing is not possible.
D) When circling to land and lose visual.
E) When instructed by ATC.

2. ON A NON-PRECISION APPROACH PROCEDURE HOW IS THE MISSED APPROACH POINT DETERMINED ? (IFH AC90-1A EXERPT)

The pilot normally determines MAP by timing from the final approach fix. MAP may also be determined thru use of DME or a specific fix utilizing VOR, ADF, RNAV if authorized on the approach chart.

3. THE MISSED APPROACH POINT ON A STRAIGHT IN NON-PRECISION APPROACH OCCURS WHERE GEOGRAPHICALLY ? (IFH AC90-1A EXERPT)

Over the runway threshold.

4. IF NO FINAL APPROACH FIX IS DEPICTED, HOW IS MAP DETERMINED ? (IFH AC90-1A EXERPT)

The MAP is at the airport. (navaid on airport)

5. WHERE IS THE MISSED APPROACH POINT ON A PRECISION APPROACH? (IFH AC90-1A EXERPT)

At the decision height on glideslope.

6. UNDER WHAT CONDITIONS ARE MISSED APPROACH PROCEDURES PUBLISHED ON AN APPROACH CHART NOT FOLLOWED ? (IFH AC90-1A EXERPT)

When ATC has assigned alternate missed approach instructions.

7. IF, DURING THE EXECUTION OF AN INSTRUMENT APPROACH PROCEDURE, YOU DETERMINE A MISSED APPROACH IS NECESSARY DUE TO FULL SCALE NEEDLE DEFLECTION, WHAT ACTION IS RECOMMENDED ? (AIM PARA 5-57)

Protected obstacle clearance areas for missed approach are predicated on the assumption that the abort is initiated at MAP not lower than MDA or DH. Reasonable buffers are provided for normal maneuvers. However, no consideration is given to an abnormally early turn. Therefore, when an early missed approach is executed, pilots should fly the IAP as specified on the approach plate to MAP at or above MDA or DH, before executing a turning maneuvering.

8. WHAT IS A LOW APPROACH ? (AIM PARA 4-62)

An approach over an airport or runway following an instrument approach procedure or a VFR approach including a go around maneuver where the pilot intentionally does not make contact with runway. It is used to expedite particular operations such as a series of practice approaches.

F. LANDING PROCEDURES

1. IS IT LEGAL TO LAND A CIVIL AIRCRAFT IF THE ACTUAL VISIBILITY IS BELOW THE MINIMUMS PUBLISHED ON THE APPROACH CHART ? (FAR 91.175)

No; No pilot operating an aircraft, except a military aircraft of the U.S., may land that aircraft when the flight visibility is less than the visibility prescribed in the standard instrument approach procedure being used.

2. WHEN LANDING AT AN AIRPORT WITH AN OPERATING CONTROL TOWER FOLLOWING AN IFR FLIGHT, MUST THE PILOT CALL FSS TO CLOSE THE FLIGHT PLAN ? (AIM PARA 5-13)

No; If operating on an IFR flight plan to an airport with a functioning control tower, the flight plan will automatically be closed upon landing.

3. IS THE REPORTED CEILING A REQUIREMENT FOR LANDING ? (IFH AC90-1A EXERPT)

No; An aircraft may still be in the clouds when at DH or MDA but have the runway environment in sight. In that situation a descent for landing is authorized.

G. LOGGING FLIGHT TIME

1. WHAT CONDITIONS ARE NECESSARY FOR A PILOT TO LOG INSTRUMENT TIME ? (FAR 61.51)

A pilot may log as instrument flight time only that time during which he/she operates the aircraft solely by reference to instruments, under actual or simulated flight conditions.

2. WHEN LOGGING INSTRUMENT TIME, WHAT SHOULD BE INCLUDED IN THE EACH ENTRY ? (FAR 61.51)

Each entry must include the place and type of each instrument approach completed and the name of the safety pilot if applicable.

3. WHAT ARE THE MINIMUM QUALIFICATIONS FOR THE INDIVIDUAL WHO OCCUPIES THE OTHER SEAT AS SAFETY PILOT DURING A SIMULATED INSTRUMENT FLIGHT ? (FAR 91.109)

He/she must be appropriately rated in the aircraft. (ASEL, MEL, ETC.)

H. INSTRUMENT APPROACH PROCEDURE CHARTS - GENERAL (ALL QUESTIONS REFERENCE GOVERNMENT NOAA CHARTS)

1. IF A PARTICULAR APPROACH NAME HAS A LETTER "A" ATTACHED AS A SUFFIX SUCH AS "VOR DME A" THIS INDICATES WHAT ? (IFH AC90-1A)

A letter suffix beside the approach name indicates the approach does not meet straight-in criteria and only circling minimums are available.

2. DO ALL STANDARD INSTRUMENT APPROACH PROCEDURES HAVE FINAL APPROACH FIXES ? (IFH AC90-1A)

No; Some non-precision approaches may not have a final approach fix. These particular approaches usually have the navaid upon which the approach is based located on the airport.

3. WITH NO FAF AVAILABLE, WHEN WOULD FINAL DESCENT TO THE PUBLISHED MDA BE STARTED ? (IFH AC90-1A)

When flying the full procedure, usually upon completion of the procedure turn and when established on the final approach course inbound. When being radar vectored to the final approach course descent shall be accomplished when within the specified distance from the navaid and established on the inbound course.

4. WHY WOULD AN AIRPORT, WITH A STANDARD INSTRUMENT APPROACH PROCEDURE AVAILABLE, BE DESIGNATED "NOT AUTHORIZED" AS AN ALTERNATE ?

If an airport is "Not Authorized" as an alternate under any conditions one of the following usually applies :

A) The Navaid upon which the approach is based is not monitored by an ATC facility; The navaid could malfunction and possibly shut down without ATC being immediately of aware of condition.

B) The airport does not have approved weather reporting capability.

I. INSTRUMENT APPROACH PROCEDURE CHARTS - PLANVIEW.

THE FOLLOWING QUESTIONS ARE IN REFERENCE TO THE ILS 16L APPROACH CHART FOR FORT WORTH, TEXAS DEPICTED AT THE END OF THIS GUIDE. (NOS EFFECTIVE DATE DEC.14,1989)

1. WHAT ARE THE MSA'S FOR THIS APPROACH ? (IFH AC90-1A)

2600 ..... 010 thru 270 degrees.
3400 ..... 270 thru 010 degrees.

2. THE MSA IS CENTERED ON WHICH FACILITY AND WHAT DOES IT PROVIDE ? (IFH AC90-1A EXERPT)

Mufin LOM; Provides 1000 ft clearance above highest obstacle in defined sector to 25 nautical miles.

3. WHAT IS THE IAF FOR THIS PROCEDURE ? (IFH AC90-1A)

Mufin LOM.

4. WHAT SIGNIFICANCE DOES THE BOLD ARROW EXTENDING FROM BRIDGEPORT VOR HAVE ? (IFH AC90-1A)

It represents a feeder route or flyable route utilized when transitioning from the enroute structure to the initial approach fix.

5. WHEN INTERCEPTING A LOCALIZER WITH A LOM FROM PROCEDURE TURN INBOUND, WHAT WILL THE RELATIVE BEARING ON THE ADF INDICATOR BE AS THE NEEDLE BEGINS TO CENTER ? (IFH AC90-1A EXERPT)

315 degrees.

6. WHAT ARE THE FREQUENCIES FOR THE LOCATOR OUTER MARKER AND MIDDLE MARKER BEACONS ? (IFH AC90-1A EXERPT)

The locator frequency is 365 kHz. All marker beacons transmit on a frequency of 75 MHz.

7. THE INNER RING LABELLED 10 NM AND CENTERED ON THE MUFIN LOM, HAS WHAT SIGNIFICANCE ? (IFH AC90-1A EXERPT)

The Inner Ring, normally a 10 NM radius, provides the boundary of the procedure that is charted to scale.

8. WHERE DOES THE FINAL APPROACH SEGMENT BEGIN FOR THE ILS 16L APPROACH ? (IFH AC90-1A)

On all precision approaches, the final approach segment begins when the glide slope is intercepted. For non-precision approaches such as the straight-in LOC 16L approach, the final approach segment begins at the Mufin LOM.

J. INSTRUMENT APPROACH PROCEDURE CHART - PROFILE

1. WITHIN WHAT DISTANCE FROM THE MUFIN LOM MUST THE PROCEDURE TURN BE COMPLETED ? (IFH AC90-1A EXERPT)

10 Nautical Miles.

2. IF A PROCEDURE TURN IS REQUIRED, WHAT WOULD BE THE MINIMUM ALTITUDE WHILE FLYING THIS SEGMENT ? (IFH AC90-1A)

The minimum altitude for the initial approach segment and while executing the procedure turn is 2200 ft. MSL.

3. WHAT ALTITUDE MAY A PILOT DESCEND TO AFTER THE PROCEDURE TURN ? (IFH AC90-1A EXERPT)

2000 MSL.

4. WHAT DOES THE NUMBER "1991" LOCATED AT THE OUTER MARKER INDICATE ? (IFH AC90-1A EXERPT)

The altitude of the glide slope at the outer marker.

5. WHAT IS THE GLIDE SLOPE ANGLE FOR THIS APPROACH ? (IFH AC90-1A EXERPT)

3.00 DEGREES.

6. WHAT IS THE ALTITUDE AT WHICH THE ELECTRONIC GLIDE SLOPE CROSSES THE THRESHOLD OF RUNWAY 16L ? (IFH AC90-1A)

57 ft.

7. IF THE GLIDE SLOPE BECAME INOPERATIVE, COULD YOU CONTINUE THIS APPROACH IF ESTABLISHED ON LOCALIZER AT THE TIME OF MALFUNCTION ? WHY? (IFH AC90-1A EXERPT)

Yes, provided ATC is notified and approves localizer only approach. The procedure indicates a localizer only minimum, meaning that a localizer only approach can be authorized. The minimum is now an MDA and the approach is now a non-precision procedure with MAP being a time or DME point.

8. IF YOU DISCOVERED YOUR MARKER BEACON RECEIVER WAS INOPERATIVE WHAT ARE THE AUTHORIZED SUBSTITUTES FOR MUFIN OUTER MARKER ? (IFH AC90-1A)

A) The Compass Locator (365 Khz)
B) 5.3 DME I-FTW
C) Radial 267 DFW VORTAC

9. WHAT DME DISTANCE IS INDICATED IN THE PROFILE VIEW FOR THE MUFIN LOM AND THE RUNWAY THRESH OLD ? (IFH AC90-1A)

The Mufin LOM is 5.3 NM amd the runway threshold is 1.5 NM from the localizer antenna site .

10. WHERE IS THE MAP FOR THE PRECISION AND NON- PRECISION APPROACH IN THIS PROCEDURE ? (IFH AC90-1A EXERPT)

Upon reaching the DH of 910 on the glide slope for the precision procedure.

For the non-precision procedure:

A) D1.5 from IFTW or
B) MUFIN to MAP D3.8 or
C) Time from MUFIN.

K. INSTRUMENT APPROACH PROCEDURE CHART - MINIMUMS SECTION

1. **WHAT IS THE MINIMUM VISIBILITY FOR A CATEGORY A FULL ILS 16L APPROACH ? (IFH AC90-1A EXERPT)**

   1/2 Mile or 2400 RVR.

2. **IF THE APPROACH LIGHT SYSTEM BECAME INOPERATIVE, HOW WOULD YOU DETERMINE THE MINIMUM VISIBILITY FOR A CATEGORY A 16L FULL ILS APPROACH ? (IFH AC90-1A EXERPT)**

   To determine landing minimums when components or aids of the system are inoperative or are not utilized, inoperative components or visual aids tables are published and normally appear in the front section of NOAA approach chart books.

3. **IS THE RVR READING SLANT RANGE, OR HORIZONTAL ? (AIM PARA 512)**

   Horizontal.

4. **ARE TAKEOFF MINIMUMS STANDARD OR NON-STANDARD FOR FT. WORTH MEACHAM FIELD ? (IFH AC90-1A EXERPT)**

   They are non-standard or a departure procedure has been published as noted by the symbol shown under the minimums box indicating that a separate listing should be consulted.

5. **FOR THE STRAIGHT-IN LOCALIZER APPROACH 16L, WHAT ARE THE MINIMUMS FOR A CATEGORY A AIRPLANE IF A CIRCLING MANEUVER IS DESIRED ? (IFH AC90-1A)**

   The circling MDA is the same as the straight-in MDA, 1260 MSL. The visibility requirement increases for the circling maneuver, going up to 1 mile.

6. WHAT SIGNIFICANCE DO THE NUMBERS IN PARENTHESES (200 1/2) HAVE ? (IFH AC90-1A EXERPT)

All minimums found in parentheses are not applicable to Civil pilots. These minimums are directed at military pilots who should refer to appropriate regulations.

7. WHEN ESTABLISHED AT THE MDA ON THE FINAL APPROACH COURSE INBOUND FOR THE STRAIGHT-IN LOC 16L APPROACH, THE MDA IS EXPRESSED AS HAT OR HAA ? (IFH AC90-1A EXERPT)

550 feet HAT; The MDA for a straight-in landing is always expressed as height above touchdown (HAT) since the approach is for a specific runway. MDA's for circling approaches will always be height of airport (HAA) since a specific runway will not be used for landing.

8. WHAT CRITERIA ARE APPROACH CATEGORIES BASED ON ? (IFH AC90-1A EXERPT)

1.3 x Vso at maximum landing weight and forward CG.

9. IF THE CURRENT WEATHER REPORTS INDICATE CEILINGS 100 OVERCAST AND VISIBILITY 1/2 MILE, CAN A PILOT LEGALLY MAKE AN APPROACH TO ILS 16L, AND CAN HE LAND ? (FAR 91.175)

Under FAR part 91 regs., the approach may be attempted regardless of the ceiling and visibility; At the decision height the pilot must have the runway environment in sight and have the prescribed flight visibility to land. If these conditions are met, the approach may be continued to a landing.

L. INSTRUMENT APPROACH PROCEDURE CHART - AERODROME.

1. WHAT TYPES OF LIGHTING ARE AVAILABLE FOR RUNWAY 16L ? (IFH AC90-1A EXERPT)

HIRL - High intensity runway lighting
MALSR - Medium intensity approach lighting system with sequenced flashing lights

2. WHAT IS THE TOUCHDOWN ZONE ELEVATION FOR RUNWAY 16L ? (IFH AC90-1A EXERPT)

710 MSL.

3. WHAT IS THE BEARING AND DISTANCE OF THE MISSED APPROACH POINT FROM THE FAF ? (IFH AC90-1A EXERPT)

164 degrees 3.8 NM from FAF.

ILS RWY 16L

AL-159 (FAA)

FORT WORTH MEACHAM (FTW)
FORT WORTH, TEXAS

REGIONAL APP CON
118.1 379.9 NORTH
120.5 227.8 SOUTH
MEACHAM TOWER★
118.3 (CTAF) 257.8
GND CON
121.9
CLNC DEL
124.65
ATIS★ 120.7
FORT WORTH RADIO
122.6

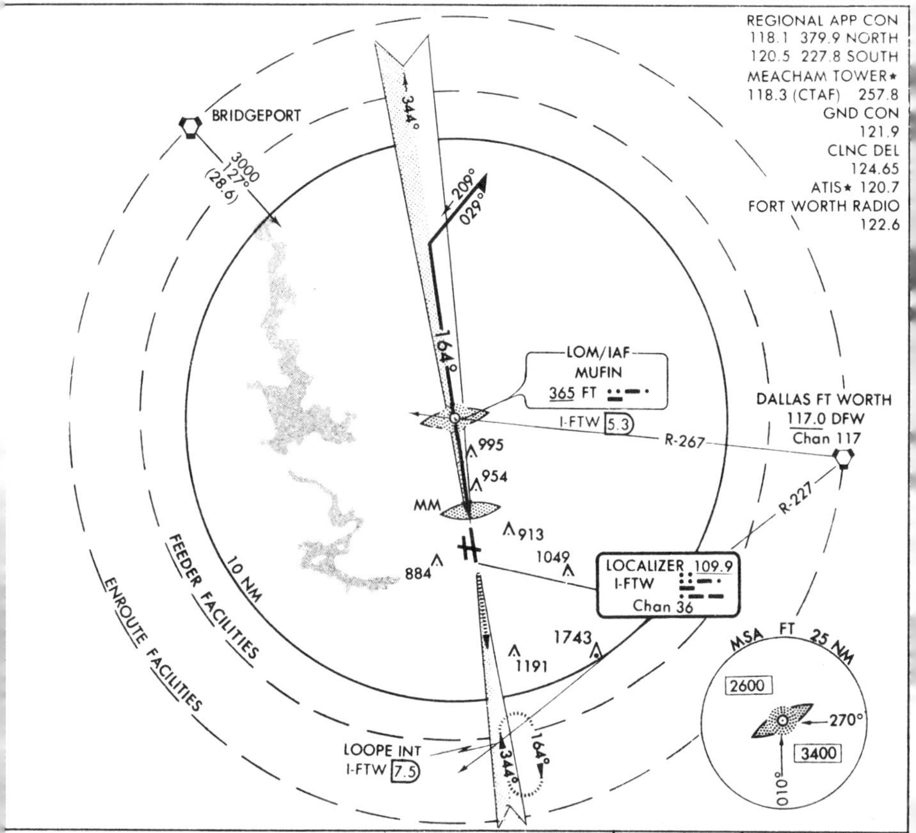

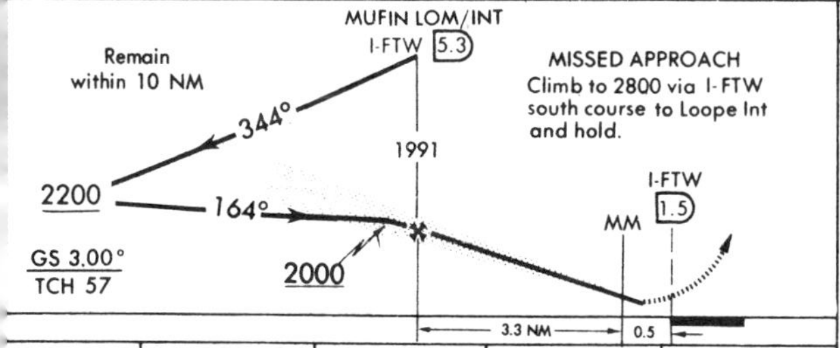

CATEGORY	A	B	C	D
S-ILS 16L	910/24 200 (200-½)			
S-LOC 16L	1260/24 550 (600-½)		1260/50 550 (600-1)	1260/60 550 (600-1¼)
CIRCLING	1260-1 550 (600-1)		1280-1½ 570 (600-1½)	1300-2 590 (600-2)

ELEV 710

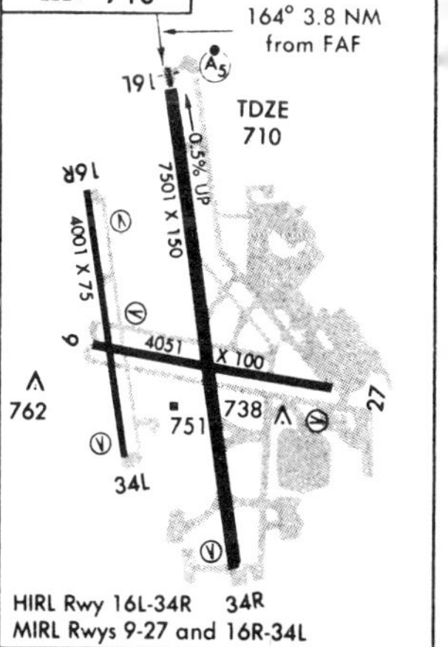

HIRL Rwy 16L-34R
MIRL Rwys 9-27 and 16R-34L

FAF to MAP 3.8 NM					
Knots	60	90	120	150	180
Min:Sec	3:48	2:32	1:54	1:31	1:16

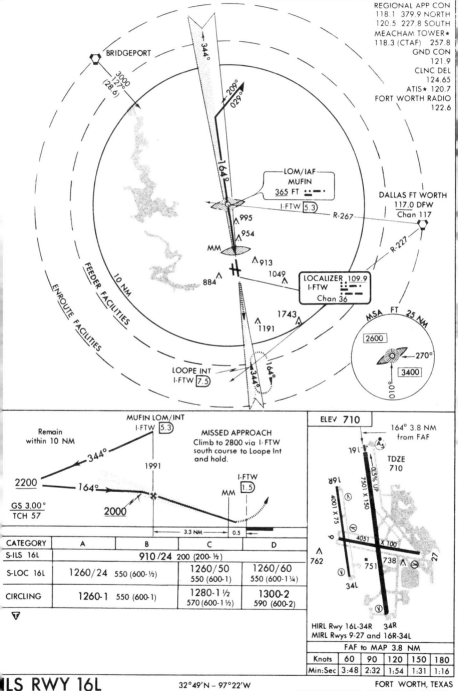

NOT FOR NAVIGATIONAL USE

NOTES

NOTES

NOTES

NOTES

NOTES

NOTES

NOTES